AF556063

Agricultural Development in Punjab and Haryana

Achievements and Challenges

Agricultural Development in Punjab and Haryana

Achievements and Challenges

Varinder Sharma

First Published 2019

ISBN 978-93-83723-52-2

Published by
LG PUBLISHERS DISTRIBUTORS
49, Street No. 14, Pratap Nagar,
Mayur Vihar Phase I, Delhi 110 091
Email: lgpdist@gmail.com

Laser Typeset at
Sakshi Computers, Delhi

Printed at
D.K. Fine Art Press, Delhi

Contents

List of Tables

List of Figures

Preface

British India was partitioned into India and Pakistan in August 1947. With this partition, East Punjab became a part of India and West Punjab integrated with Pakistan. Again in 1966, East Punjab on the basis of language and geography reorganised into Punjab and Haryana. The hilly area of the state carved into the present state of Himachal Pradesh.

On the eve of partition the agriculture and industrial development of East Punjab was completely in the doldrums. In front of planners and policy makers the issue of food security of India was a main concern. At this critical juncture, Punjab and Haryana turned into harbingers of agricultural development in India. With the slogan of "Green Revolution" these two states became a role model of advancement in agriculture in terms of production and productivity for other states. Gradually, the two states became main contributors to the central pool of food grains. The two main food grain crops of these states are rice and wheat.

Within these decades since the period of the beginning of the green revolution, the main changes came into cropping patterns and agriculture infrastructure, etc. of these two states. Moreover, the development economists, leftist scholars and social activists also raised their voices against the unequal distribution created by this new agricultural development between landless and landowners.

No such work exists which critically compared the process of agricultural development in these two states. The present work tries to bridge the gap in research.

Acknowledgements

The idea to pursue this present study occurred to me long ago. The idea could not materialise owing to lack of motivation and laziness. Long back this idea was discussed with Prof. H.S. Shergill and Prof. Lakhwinder Singh Gill, Chairperson, Department of Economics, Punjabi University, Patiala. Both economists motivated me to undertake this study. They not only gave me motivation but immense guidance. I sincerely acknowledge them.

I am also thankful to Prof. M.R. Khurana, for giving me training to write a script in the form of a book and guidance regarding this study. The discussion with Prof. Jaswinder Brar, Department of Economics, Punjabi University, Patiala on current agrarian issues of Punjab and Haryana always remained useful. I sincerely acknowledge him.

My special thanks to Prof. Pramod Kumar, Director, Institute for Development and Communication (IDC), Chandigarh for providing ample time and facilities to finish this document. The discussion with him always remained useful to conceptualise the context.

Finally, I acknowledge the farmers of Punjab and Haryana for clarifying certain current issues in agriculture.

The contribution of Ms. Dilpreet Kaur, Research Scholar and Ms. Shivangi Mittal, Intern at IDC from the Department of Economics, Panjab University, Chandigarh is acknowledged for collecting data from various publications. Mr. Amit Kumar, Research Scholar at IDC helped to design the maps used in this study.

Mr. Narinder Kumar, typed this manuscript many times meticulously for which I acknowledge his patience.

Author

Acronyms and Abbreviations

CBB	:	Commercial Bank Branch
HYVs	:	High-Yielding Seed Varieties
IDC	:	Institute for Development and Communication
Kgs.	:	Kilograms
Kwh	:	Kilowatt
LDB	:	Land Development Bank
NSDP	:	Net State Domestic Product
PACS	:	Primary Agriculture Cooperative Society
RRBB	:	Regional Rural Bank Branch
SYL	:	Satluj Yamuna Link Canal
USA	:	United States of America
UT	:	Union Territory
WB	:	World Bank

CHAPTER 1

Introduction

The name Punjab is derived from the name of five rivers (the Jhelum, Ravi, Beas, Satluj and Chenab). These tributaries flow out of the Himalayan Mountains and combine to form the Indus River[1]. During British rule, Punjab was comprised of the north-western and Indo-Gangetic plain. It was bounded on the east by the Yamuna River, on the north and west by the two mountain ranges, viz. the Himalayas and Sulaiman, in the South by the Thar Desert[2]. The Mughal Empire declined in the eighteenth century, a Sikh Kingdom established in the Punjab and remained independent until 1849, and then this entire region was annexed by the British.

With the partition of India, into India and Pakistan in August 1947, western Punjab became part of Pakistan and eastern Punjab is now the present Indian Punjab.

The partitioned portion of India inherited about 47 per cent of the population, 34 per cent of the area and 20 per cent of the canal irrigated area. Moreover, it was a grain deficit area[3]. Most of the Hindu and Sikh population of western Punjab (4.3 million) migrated into India while most of the Muslim population of eastern Punjab (4.2 million) crossed over to Pakistan[4]. By 1951, 2.73 million refugees from Pakistan had settled in Indian Punjab and comprised 17 per cent of the state's population[5]. Around 37 per cent of these migrated families were not involved directly into cultivation and settled primarily in the towns and cities[6].

Again in November 1966, the state of Punjab was bifurcated on the basis of language into the present states of Punjab and Haryana and some hill portion of the state was added into a state of Himachal Pradesh[7]. The division on the basis of language gave Punjab a majority of Sikh and Punjabi speakers, whereas in Haryana, Hindus and Hindi speakers dominate[8]. The states have a common capital, Chandigarh (UT). The location of two states and its capital city in India is depicted in Map 1.1 and administrative divisions of these states in Map 1.2 and Map 1.3 [See Appendix-A].

After the re-organisation of the state, the period of regeneration and development started. The state governments formulated and executed the development plans with particular strategies and priorities. The prime task before the state governments was to end food deficit. A large tract in two states had been suffering from droughts due to shortage of irrigation and inadequate rainfall. Therefore plans were formulated to insulate this tract against the vicissitudes of drought.

The economic conditions in India during the mid-1960s, on the eve of introduction of high-yielding seed varieties (HYVs), were the worst ever during the post-independence period. The growth of per capita income reached its low water mark and major industries were badly hit by recession and unemployment was mounting.

The country was not producing sufficient output of food grains to feed the rising population. It was heavily dependent on imports of food grains from the USA. At the same time, it was always uncertain about the food surplus countries in the world to continuously supply the needs of the food-deficient nations.

In India's case this uncertainty of supply of foodgrains from food surplus nations to food deficient countries turned into horror when the Paddock brothers'[9] thesis came out by 1975. Their thesis predicted widespread famine in various parts of the world where gaps prevailed in supply and demand of foodgrains. At that time only the USA was the food surplus country. A policy of discrimination in favour of a particular food-deficit country could only save a country from famine. The Paddock brothers considered India a hopeless case and hypothesised that it could not be saved from starvation. Their thesis sounded an alarm and drew the attention of policy makers. It was a debatable issue that supply and demand conditions in the world food market were against India.

At this critical juncture, therefore, virtually there was no option except to seek self-sufficiency in foodgrain production.

The increase in foodgrain production was not possible by increasing the area under cultivation but it was feasible by raising the productivity of land. Therefore, during the mid-1960s the Borlaug[10] seed fertiliser technology was adopted in selected parts of India. Both Punjab and Haryana became leading states in the production of foodgrains within four years from the inception of new agriculture technology, these states received the worldwide fame for the success of the green revolution.

At present, Punjab and Haryana have around 2.91 and 2.49 per cent share in the net sown area of India. But these two states have sufficient potential of foodgrains to feed the Indian masses. These two states contribute 53 per cent of rice and 75 per cent of wheat to the central pool of food grains. The country at present is in a position to even export some food grains especially rice. Thus, these two states clearly falsified the prediction of the Paddock brothers' thesis about the grim position of India's foodgrains situation. But in this process of agricultural development, Punjab in terms of production of foodgrains and use of agriculture inputs remained ahead of Haryana[11]. In recent times, some scholars also tried to compare the process of agricultural development in case of other states[12]. But there is a gap in literature in the case of Punjab and Haryana.

Here in this study, the process of agricultural development in these two states has been compared from the early green revolution years to the post green revolution period. The study is arranged in the following chapters.

In Chapter 1, a brief introduction of production of foodgrains from the partition period to the green revolution period is discussed.

In Chapter 2, the scenario of agricultural development in the early green period in two states Punjab and Haryana is compared.

Chapter 3, focuses on the level of agricultural development in two states in the current period.

Chapter 4, explains how the gains of agricultural development in two states percolated among the landless agriculture labourers and small farmers.

Chapter 5, deals with the conclusions and summary.

REFERENCES

1. The geographical location and the five tributaries (Jhelum, Ravi, Beas, Satluj and Chenab) are discussed by Richard P. Day and Inderjit Singh in *Economic Development as an Adaptive Process: The Green Revolution in the Indian Punjab*, 1977, p. 44.
2. For more details on the evolution of boundaries of colonial Punjab, see Joseph E. Schwartzberg (edt), *A Historical Atlas of South Asia*, 1978.
3. See, *Techno Economic Survey of Punjab*, National Council of Applied Economic Research (NCAER), 1962, p. 105.
4. M.S. Randhawa, *Green Revolution: A Case Study of Punjab*, 1974, p. 30.
5. See, Kusum Nair, *Blossoms in the Dust*, 1961, p. 113.
6. For details, see M.L. Pandit, "Some Less Known Factors Behind Recent Industrial Change in Punjab and Haryana", *Economic and Political Weekly*, Vol. 13, No. 17, 1978, p. 1937.
7. See A.S. Narang: "Punjab: Development and Politics" in *Land, Caste and Politics in Indian States* (ed.), 1982, p. 118.
8. Ibid
9. Paddock, William and Paddock, Paul (1967): *Famine 1975! America's Decision: Who Will Survive?* Little Brown & Company, Georgia, USA.
10. Norman Ernest Borlaug was an American agronomist who developed high-yielding disease-resistant wheat varieties. Later on these high-yielding varieties with advanced agricultural production techniques increased agriculture production and productivity in India, Pakistan and Mexico. For details, see Wikipedia.
11. See Appendix-33 and Serial Numbers 28, 39, 45, 46 and 47.
12. For the inter-state comparison of development see Arvind Panagariya and M. Govind Rao (eds.): *The Making of Miracles in Indian States: Andhra Pradesh, Bihar and Gujarat*, 2015.

CHAPTER 2

Agricultural Development in Punjab and Haryana on the Eve of the Green Revolution Period

In these two states, the process of agricultural development was initiated during the period of the 1960s onwards by huge public investment especially in irrigation, agricultural research and extension services. The states of Punjab and Haryana adopted the new agriculture technology during the 1960s. In this chapter, the level of agricultural development in the two states, viz. Punjab and Haryana is discussed in the early phase of the green revolution.

Classification of Area

The classification of area as per village records is given in Table 2.1. The state of Punjab seems to have a larger tract of total area than the state of Haryana. The triennial averages from 1965-66 to 1967-68 (Table 2.1 and Figure 2.1) elaborate the total area of Punjab which was marginally more by 627 thousand hectares compared to Haryana.

The total cropped area in Punjab was 5164 thousand hectares, whereas in Haryana, it touched 4605 thousand hectares.

The net sown area in Punjab was 465 thousand hectares more than Haryana. Further, if we look at Table 2.1 and Figure 2.1, it visualises that the area under forests and fallow lands was marginally more in Haryana than Punjab. But the uncultivated area exceeded in Punjab than Haryana. It means in Punjab during the 1960s there was still some scope to bring more area under cultivation.

Table 2.1

Utilisation of Area in Punjab and Haryana (Average Over 1965-66 to 1967-68) ('000 Hectares)

Classification of Land Area	Punjab (i)	Percent-age	Haryana (ii)	Percent-age	Difference (i)-(ii)
Total Cropped Area	5164	-	4605	-	559
Cropping Intensity	133	-	134	-	-
Net Sown Area	3891	77.42	3426	77.88	465
Uncultivated Area	815	16.22	627	14.25	188
Forests	82	1.63	90	2.05	-8
Fallow Land	238	4.74	256	5.82	-18
Total Area	5026	100.00	4399	100.00	627

Source: *Statistical Abstracts of Punjab and Haryana* for Various Years.
Note: Percentages are Out of Total Area.

Figure 2.1

Classification of Area in Punjab and Haryana ('000 Hectares) During the 1960s

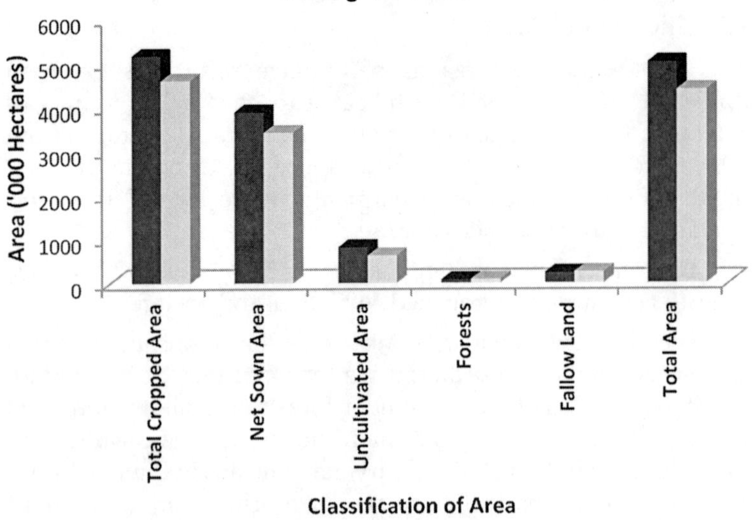

Structure of Agricultural Workforce

The rural population during the decade of 1951 to 1961, increased in Punjab by 1.89 per cent per annum and in Haryana by 2.93 per cent[1]. This growth in the rural population had also a significant impact on rural labour markets. The number of agricultural labourers and cultivators also continuously swelled in Punjab and Haryana. The total agriculture workforce is described in Table 2.2.

Table 2.2

Agriculture Workforce in Punjab and Haryana (1961) (lakh)

Categories	Punjab (i)	Percentage	Haryana (ii)	Percentage	Difference (i)-(ii)
Agricultural Labourers	3.34	17.25	1.99	9.77	1.35
Cultivators	16.02	82.75	18.38	90.23	-2.36
Agricultural Workers	19.36	100.00	20.37	100.00	-1.01

Source: *Statistical Abstracts of Punjab and Haryana* for Various Years.

In the early 1960s the total agriculture workers (agricultural labourers and cultivators) were 20.37 lakhs in Haryana and 19.36 lakhs in Punjab. But the number of agricultural labourers was 3.34 lakhs in Punjab, whereas in Haryana it was 1.99 lakhs. This difference was due to the greater use of hired labour in Punjab agriculture compared to Haryana. Seventy per cent of the total agricultural labourers were from Scheduled Castes in Punjab during the 60s[2]. Conversely to this the number of cultivators were more in Haryana (18.38 lakhs) and in Punjab this figure touched 16.02 lakhs only. Besides, these two categories of agriculture workers during the 60s, other categories were sharecroppers, biswedars and share wage servants in both states[3].

Share of Primary Sector in Net State Domestic Product (NSDP)

In the developing economies in the process of economic growth and industrialisation, the share of the primary and agricultural sector is always large[4]. In the early years of the green revolution, the contribution of agriculture and animal husbandry was highest in the total net state domestic product. In this period in both states the share of agriculture and animal husbandry was around fifty per cent.

Table 2.3
Percentage Share of Primary Sector in Net State Domestic Product (At Constant Prices) (1965-66)

Industry	Punjab (i)	Haryana (ii)	Difference (i)-(ii)
Agriculture and Animal Husbandry	50.6	54.68	-4.08
Forestry	0.22	0.12	0.1
Fishery	0.03	0.02	0.01
Total Primary Sector	50.85	54.82	-3.97

Source: *Statistical Abstracts of Punjab and Haryana* for Various Years.

The contribution of other sub-sectors of the primary sectors, i.e. of forestry, fishery and mining was negligible. Overall contribution of the primary sector in total net state domestic product was around 50.85 per cent in Punjab and 54.82 per cent in Haryana.

Net Irrigated Area and Sources of Irrigation

In the colonial period in Punjab, the irrigated area under canals was increased to cope with the frequent famines[5]. The canal based irrigation required minimum supervisory cost and capital[6]. With the green revolution, there was need of adequate irrigation facilities for the successful cultivation of HYV seeds. A net work of canals was the main source of irrigation in Punjab in the pre-green revolution period. But most of these canals were in West Punjab and went to Pakistan after partition. Again prior to 1966, the state governments emphasised mainly canal irrigation. In Punjab and Haryana a landmark in canal irrigation came with the construction of Bhakra Dam. This dam was commissioned in 1954. In India for the first time water in canals was regulated from the dam reservoir[7]. The course of the main canals network in Punjab and Haryana is indicated in the Map (2.1) and (2.2) in the Appendix (Part A). The other irrigation project between two states was SYL (Satluj Yamuna Link Canal) which had the target to irrigate 4.46 lakh hectares in Haryana and 1.21 lakh hectares in Punjab. But this project is in the doldrums because the two states could not agree on the sharing of water of this canal (Ministry of Water Resources, River Development and Ganga Rejuvenation, Government of India).

The level of irrigation at the beginning of the green revolution was quite high in Punjab compared to Haryana. The net irrigated area in Punjab was almost two times more than Haryana. The percentage of net irrigated area to net sown area was more than fifty per cent in Punjab, whereas in Haryana, it was just 35 per cent (Figure 2.2). The percentage of gross irrigated area to gross cropped area was two times more in Punjab than Haryana (Table 2.4).

Table 2.4

Net Irrigated Area in Punjab and Haryana (Average Over 1965-66 to 1967-68) ('000 Hectares)

Level of Irrigation	Punjab (i)	Haryana (ii)	Difference (i)-(ii)
Net Irrigated Area	2276	1211	1065
Area Irrigated by Canals	1287	934	353
Area Irrigated by Tubewells	686	235	451
Percentage of Net Irrigated Area to Net Sown Area	58.49	35.33	23.16
Percentage of Gross Irrigated to Gross Cropped Area	64.32	36.17	28.15

Source: *Statistical Abstracts of Punjab and Haryana* for Various Years.

According to Table 2.5, that irrigation was more advanced in Punjab. The maximum irrigated area was under canals in both states. In Punjab almost 1287 thousand hectares of cultivated area was under canal irrigation, whereas in Haryana, it was just 934 thousand hectares. The tubewell irrigated area was 686 thousand hectares in Punjab which was three times more than Haryana. In percentage terms in Punjab, 41 per cent of the net irrigated area was irrigated by tubewells and 56.55 per cent by canals (Figure 2.3A).

Table 2.5

Net Irrigated Area by Different Sources in Punjab and Haryana (Average Over 1965-66 to 1967-68) ('000 Hectares)

Sources	Punjab (i)	Percentage	Haryana (ii)	Percentage	Difference (i)-(ii)
Canals	1287	56.55	946	77.73	341
Tubewells	952	41.83	247	20.3	705
Other Sources	37	1.62	24	1.97	13
Total Irrigated Area	2276	100.00	1217	100.00	1059

Source: *Statistical Abstracts of Punjab and Haryana* for Various Years.

Figure 2.2

Percentage of Irrigated Area in Punjab and Haryana During the 1960s

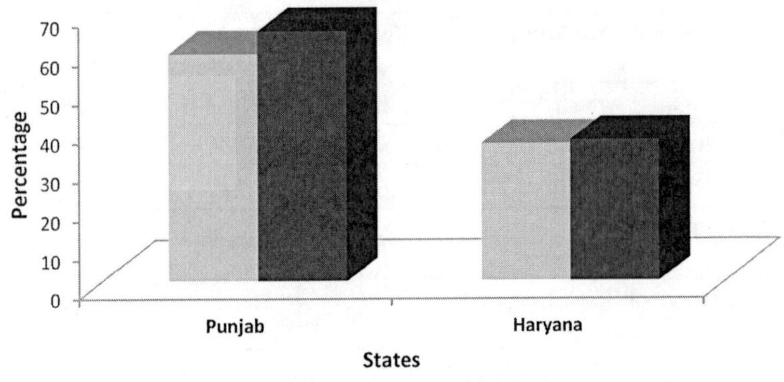

Similarly, in Haryana, the percentage of net irrigated area by tubewells was 20.30 per cent and by canals 77.73 per cent (Figure 2.3B).

Figure 2.3A

Percentage of Net Irrigated Area by Different Sources in Punjab During the 1960s

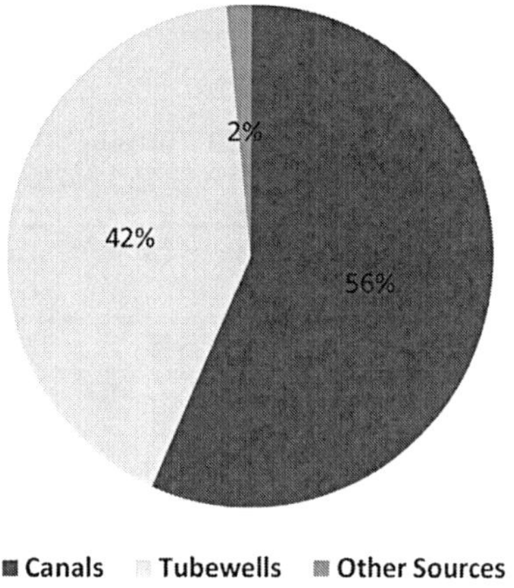

Figure 2.3B

Percentage of Net Irrigated Area by Different Sources in Haryana During the 1960s

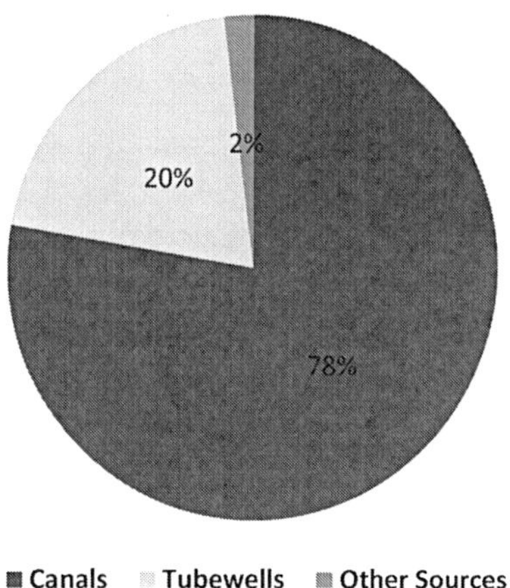

Use of Main Agriculture Inputs and Markets

The new high-yielding varieties of seeds are more responsive to controlled dosages of micro-nutrients and timely irrigation. Here we compare the use of main agriculture inputs in Punjab and Haryana. In Punjab per hectare use of fertiliser was 72 kgs, whereas in Haryana, it was just half of Punjab. In addition to this, the number of tubewells was three times more in Punjab than Haryana.

Table 2.6

Main Agriculture Inputs in Punjab and Haryana (Average Over 1965-66 to 1967-68)

Inputs	Punjab (i)	Haryana (ii)	Difference (i)-(ii)
Fertiliser ('000 Kg)	282548	82480	200068
Tubewells Used for Irrigation (Number)	33817	11003	22814
Consumption of Electricity (Lakh KWH)	2074	995	1079
Number of Agriculture Markets	84	-	-
Loans Advanced by Primary Agriculture Credit Societies (Rs. Crores)	25.58	9.89	15.69
Number of Tractors	10636	4847	5789

Source: *Statistical Abstracts of Punjab and Haryana* for Various Years.

The irrigated area under per tubewell was approximately 67 hectares in Punjab, whereas in Haryana, it was 110 hectares. The total consumption of electricity in agriculture was two and a half times more in agriculture in Punjab than Haryana.

For the effective marketing of agricultural produce especially the grains, the government developed the regulated agriculture markets. From Table 2.6, it appears in Haryana there was no regulated agriculture market at this time, but in Punjab, there were 84 regulated agriculture markets (Table 2.6).

Further, the flow of credit from the cooperatives in agriculture is given in Table 2.6. The credit cooperatives continuously developed in Punjab from the pre-independence period and during the 1950s the state governments actively supported these cooperatives[8]. Upto 1964, almost every village in Punjab had its own credit cooperative society. During 1965-66, approximately 98 per cent of all cultivators' households were the members of these societies[9]. The credit cooperatives mainly provided short-term credit to the cultivators. The state of Punjab also established the Punjab State Cooperative Land Mortgage Bank in 1958 for financing long-term cooperative credit[10].

In Punjab these credit cooperatives advanced loans of Rs. 25.58 crore to farmers during the 1960s and in Haryana the quantum of credit was only Rs. 9.89 crore. Per cultivator a loan of Rs. 150 was advanced in Punjab which was three times more than Haryana.

The other feature of the modernisation of agriculture with the green revolution was mechanisation of various farm operations. During the 1960s in Punjab there were around ten thousand tractors, but in Haryana there were just around four thousand tractors.

Lastly, the expenditure by state governments on agricultural research was also the highest in Punjab and Haryana in comparison to other states. For example, this expenditure was 75 per cent higher in Punjab and 50 per cent higher in Haryana than Gujarat[11].

Area Under Foodgrain Crops

The state of Punjab experienced impressive growth in agricultural output since the beginning of the planning era[12]. The state was reorganised in 1966 and with this the state of Haryana came into existence. During the mid-1960s, the high-yielding varieties of some major crops were adopted in agriculture and this period became the dividing line in the growth profile of agriculture of two states. In the two states during the early periods of the 60s the main crops were wheat, maize, gram and bajra. Out of the total food grain area, around fifty per cent of the area was under wheat cultivation in Punjab [Table 2.7 and Figure 2.4(A)]. In the next chapter, we have analysed the growth in area, production and yield of food grains and non-foodgrains with the linear trend models.

The remaining 17 per cent of the area was under gram and 13 per cent under maize. The remaining thirty per cent of the area was under other food grain crops. In Haryana, approximately 29 per cent of the area was under gram which was 12 per cent more than Punjab [Table 2.7 and Figure 2.4(B)].

Table 2.7

Area Under Main Foodgrain Crops in Punjab and Haryana (Average Over 1965-66 to 1967-68) ('000 Hectares)

Crops	Punjab (i)	Percent-age	Haryana (ii)	Percent-age	Difference (i)-(ii)
Rice	297	8.94	201	5.79	96
Wheat	1649	49.64	754	21.7	895
Maize	435	13.09	97	2.79	338
Bajra	183	5.51	853	24.55	-670
Jowar	5	0.15	270	7.77	-265
Barley	107	3.22	201	5.79	-94
Gram	589	17.73	1030	29.65	-441
Other Foodgrains	57	1.72	68	1.96	-11
Total Area Under Foodgrains	3322	100.00	3474	100.00	-152

Source: *Statistical Abstracts of Punjab and Haryana* for Various Years.

Figure 2.4A

Percentage of Area Under Main Foodgrain Crops in Punjab During the 1960s

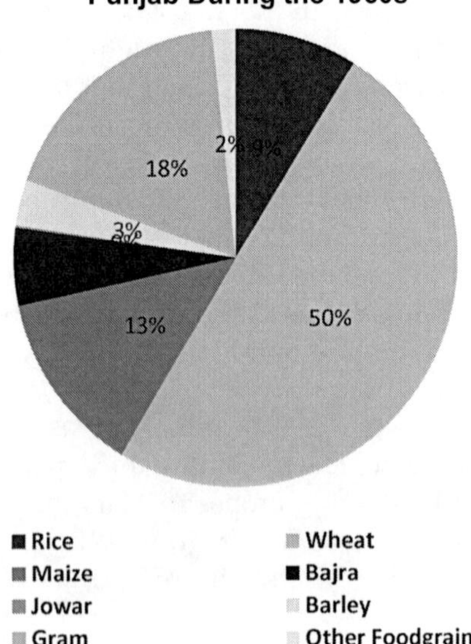

Figure 2.4B

Percentage of Area Under Main Foodgrain Crops in Haryana During the 1960s

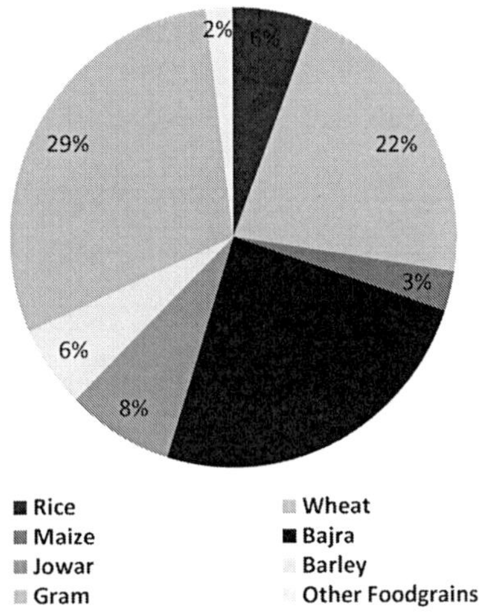

Similarly, bajra was cultivated in Haryana on more than seven times area relatively to Punjab. The crops of maize and wheat were cultivated on more area in Punjab. The rice crop did not seem an important crop during the 1960s in both states. In these states around 200 thousand hectares of area was under rice cultivation. It seems during early period of green revolution only those crops were grown which consumed less water and were rain fed.

Production of Main Foodgrains Crops

Punjab was leading the other states in terms of foodgrains production and productivity and was followed by Mysore, Maharashtra and Andhra Pradesh as regarding the growth of non-foodgrain production and productivity[13]. Thus it is clear that the state did not enjoy an all round edge over other states till the mid-1960s. Here, the production of main foodgrain crops has been elaborated and compared in Punjab and Haryana. In total foodgrain production, the share of wheat was 59.50 per cent and of maize share was just 15.69 per cent (Table 2.8 and Figure 2.5).

Table 2.8

Production of Main Foodgrain Crops in Punjab and Haryana (Average Over 1965-66 to 1967-68) ('000 Metric Tonnes)

Crops	Punjab (i)	Percent-age	Haryana (ii)	Percent-age	Difference (i)-(ii)
Rice	349	8.09	238	8.39	111
Wheat	2567	59.50	1122	39.53	1445
Maize	677	15.69	106	3.74	571
Bajra	146	3.38	347	12.23	-201
Jowar	3	0.07	47	1.66	-44
Barley	102	2.36	227	8.00	-125
Gram	444	10.29	728	25.65	-284
Other Foodgrains	26	0.60	23	0.81	3
Total Production of Foodgrains	4314	100.00	2838	100.00	1476

Source: *Statistical Abstracts of Punjab and Haryana* for Various Years.

Figure 2.5

Production of Main Foodgrain Crops in Punjab and Haryana ('000 Metric Tonnes) During the 1960s

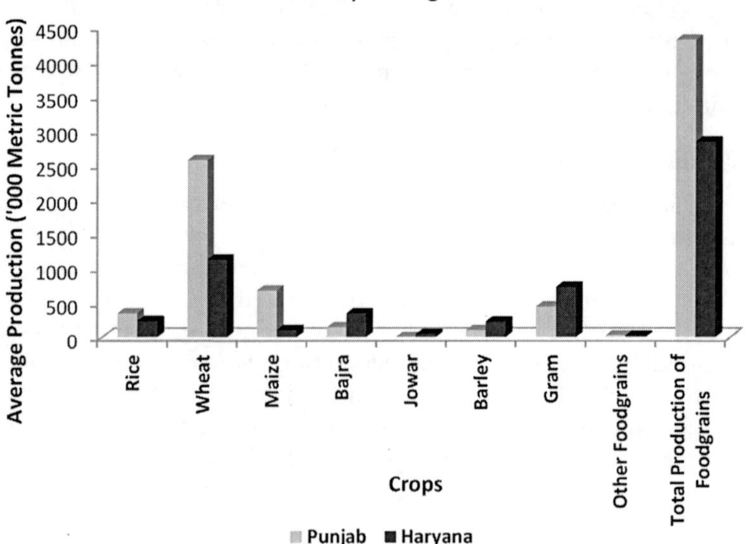

On the other side, in Haryana the maximum share was crop wheat (39.53 per cent) and gram (25.65 per cent) respectively. In Haryana, the production of maize was not so significant but the bajra crop had a share of 12.23 per cent in the total foodgrain production.

Yield of Foodgrain Crops

The yield level comparison shows, per hectare yield in kilograms is the highest in Punjab for all crops except rice and barley (Table 2.9 and Figure 2.6).

Table 2.9

Yield of Main Foodgrain Crops in Punjab and Haryana (Average Over 1965-66 to 1967-68) (Per Hectare Kgs)

Crops	Punjab (i)	Haryana (ii)	Difference (i)-(ii)
Rice	1169	1182	-13
Wheat	1542	1467	75
Maize	1560	1093	467
Bajra	786	401	385
Jowar	431	172	259
Barley	957	1193	-236
Gram	756	679	77

Source: *Statistical Abstracts of Punjab and Haryana* for Various Years.

In the case of crops like wheat and gram the gap in yield level of Punjab and Haryana is less than 100 kgs. It is clear from the previous table that maize was not cultivated on a large scale in Haryana. The yield gap in this crop between Punjab and Haryana was 400 kgs per hectare. The yield gap in case of rice in Punjab and Haryana was 13 kgs. It means that the yield was higher in Haryana than Punjab.

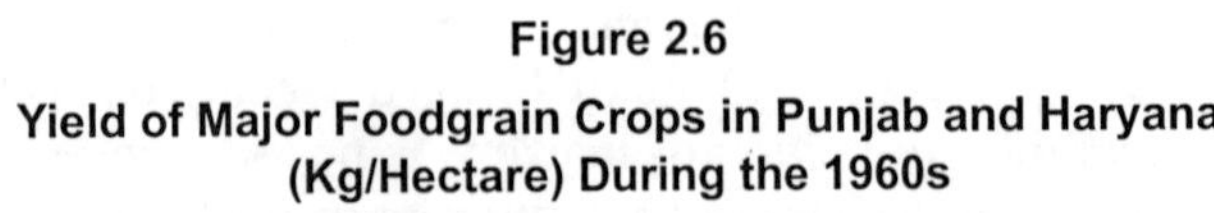

Figure 2.6

Yield of Major Foodgrain Crops in Punjab and Haryana (Kg/Hectare) During the 1960s

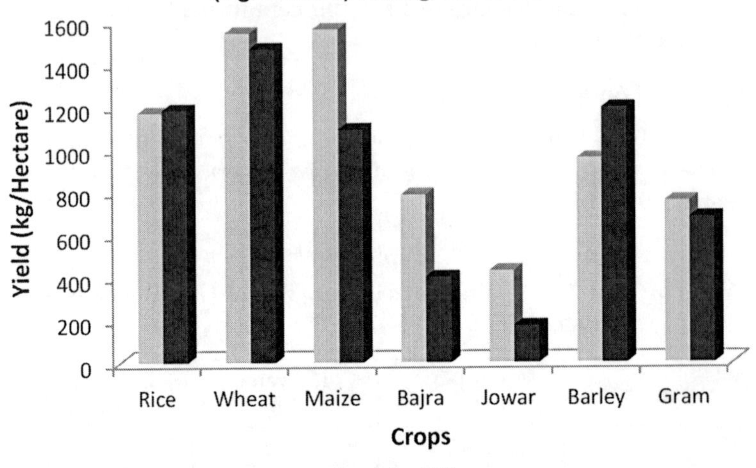

Area Under Non-Foodgrain Crops

In the early phase of the green revolution, the main non-foodgrain crops were: cotton, sugarcane, rape and mustard. In both states Punjab and Haryana, the maximum area was under cotton crop. Out of the total area under non-foodgrain crops, 48.18 per cent area in Punjab and 38.52 per cent area in Haryana was under cotton cultivation [Table 2.10 and Figures 2.7(A) & (B)].

Table 2.10

Area Under Main Non-Foodgrain Crops in Punjab and Haryana (Average Over 1965-66 to 1967-68) ('000 Hectares)

Crops	Punjab (i)	Percent-age	Haryana (ii)	Percent-age	Difference (i)-(ii)
Rape and Mustard	121	13.31	166	30.89	-45
Sugarcane	153	16.83	151	28.1	2
Cotton	438	48.18	207	38.52	231
Other Non-Foodgrains	197	21.67	13.4	2.49	183.6
Total Area Under Non-Foodgrains	909	100.00	537.4	100.00	371.6

Source: *Statistical Abstracts of Punjab and Haryana* for Various Years.

Figure 2.7A

Percentage of Area Under Main Non-Foodgrain Crops in Punjab During the 1960s

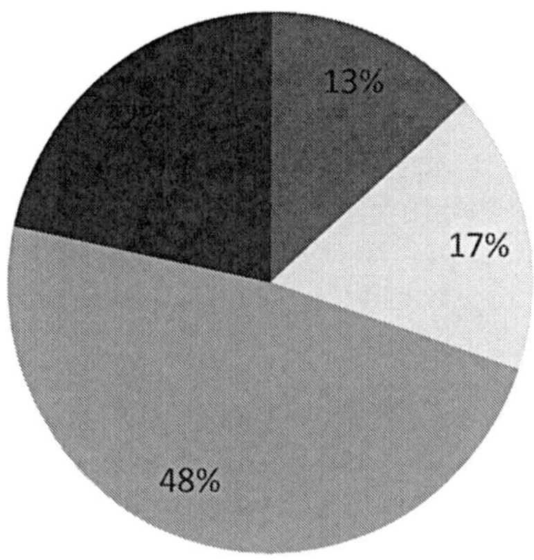

Figure 2.7B

Percentage of Area Under Main Non-Foodgrain Crops in Haryana During the 1960s

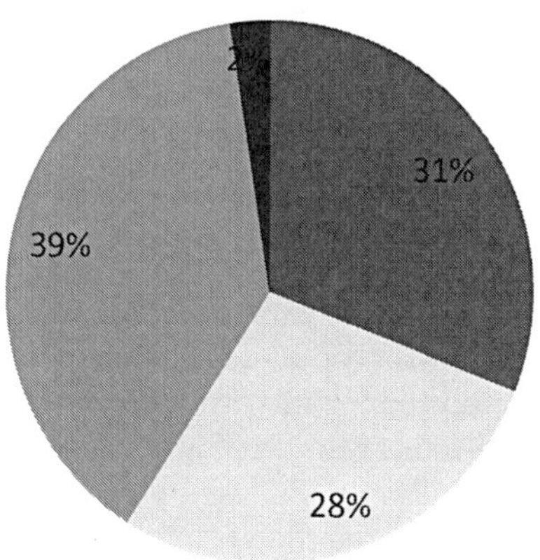

But under the other two crops sugarcane, rapeseed and mustard percentage of area was more in Haryana than Punjab. The area in Punjab under non-foodgrains is 3 times less than the total food grains and in Haryana, it is less by six times.

Production of Major Non-foodgrains Crops

In the early period of the green revolution in two states, viz. Punjab and Haryana, the maximum contribution in total non-foodgrain crops was of two crops, i.e. cotton and sugarcane. In Punjab, the share of cotton crop was 41.42 per cent and in Haryana, it was 8.39 per cent (Table 2.11 and Figure 2.8).

Table 2.11

Production of Main Non-Foodgrain Crops in Punjab and Haryana (Average Over 1965-66 to 1967-68) ('000 Tonnes)

Crops	Punjab (i)	Percentage	Haryana (ii)	Percentage	Difference (i)-(ii)
Rape and Mustard	55	4.29	43	6.33	12
Sugarcane	490	38.22	566	83.36	-76
Cotton	531	41.42	57	8.39	474
Other Non-Foodgrains	206	16.07	13	1.91	193
Total Production of Non-Foodgrains	1282	100.00	679	100.00	603

Source: *Statistical Abstracts of Punjab and Haryana* for Various Years.

On the other side, in Haryana only sugarcane was the main non-foodgrain crop. This crop had the share of more than eighty per cent in total non-foodgrain production. In both the states the other main non-foodgrain crop was rape and mustard. The production of non-foodgrains is less by almost three and a half times in Punjab and Haryana than foodgrains.

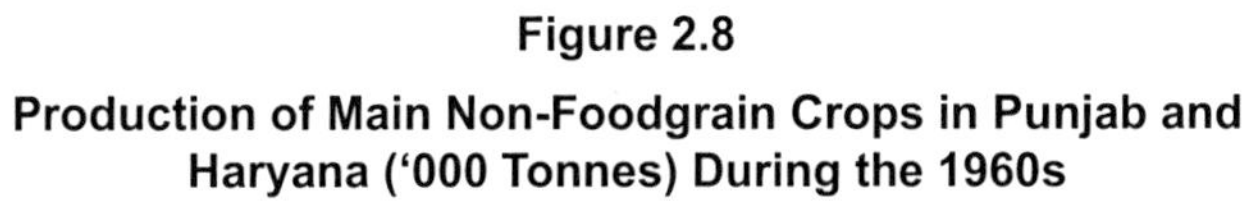

Figure 2.8

Production of Main Non-Foodgrain Crops in Punjab and Haryana ('000 Tonnes) During the 1960s

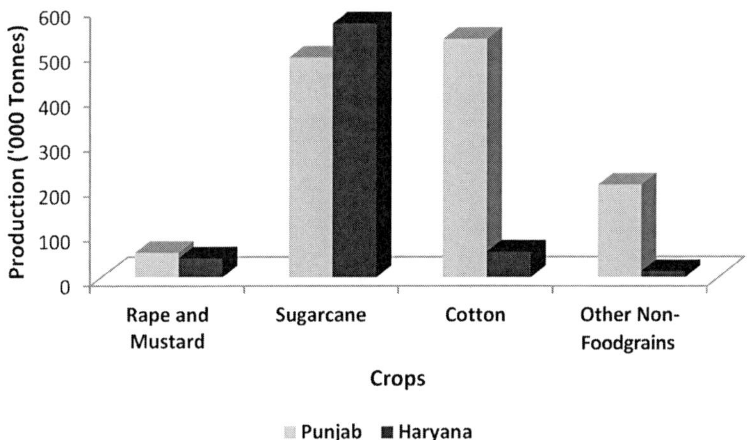

Yield of Major Non-Foodgrain Crops

After discussing the area and production of the main non-foodgrain crops in Punjab and Haryana, now, we compared the level of yield in these two states (Table 2.12 and Figure 2.9).

Table 2.12

Yield of Major Non-Foodgrain Crops in Punjab and Haryana (Average Over 1965-66 to 1967-68) (Kgs Per Hectare)

Crops	Punjab (i)	Haryana (ii)	Difference (i)-(ii)
Rapeseed and Mustard	476	412	64
Sugarcane	3255	3751	-496
Cotton	308	277	31

Source: *Statistical Abstracts of Punjab and Haryana* for Various Years.

In Punjab, per hectare yield of sugarcane was 3255 kgs per hectare, whereas in Haryana, it was 3751 kgs. On other side, the yield of rapeseed and mustard was 64 kgs per hectare more in Punjab than Haryana. The difference in the yield of cotton was of 31 kgs per hectares more in Punjab than Haryana.

Figure 2.9

Yield of Major Non-Foodgrain Crops in Punjab and Haryana (Kg/Hectare) During the 1960s

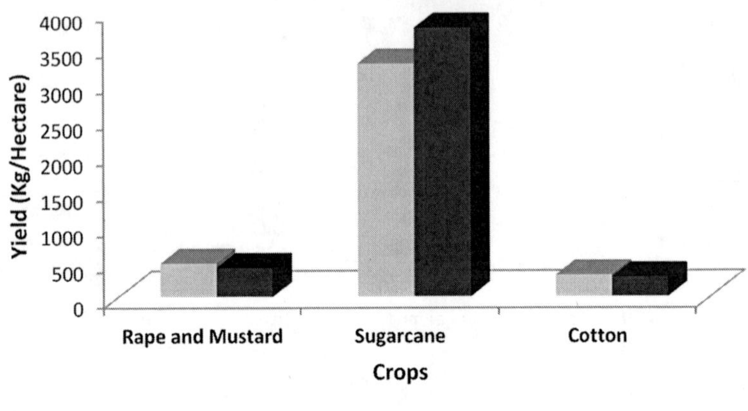

Livestock and Poultry

Since the British period, the cattle breeding was advanced in Punjab and cultivators were actively involved in keeping high quality milch animals[14]. In Table 2.13, the total number of livestock and poultry is given in Punjab and Haryana. In Punjab as well as in Haryana, the number of buffaloes was higher than cows.

Table 2.13

Livestock and Poultry in Punjab and Haryana (1966-67) (Lakhs)

Species	Punjab (i)	Percent-age	Haryana (ii)	Percent-age	Difference (i)-(ii)
Cows					
Total	8.2	100	6.1	100	2.1
In Milk	4.8	58.54	3.5	57.38	1.3
Buffaloes					
Total	29.8	100	19.3	100	10.5
In Milk	8.6	28.86	5.8	30.05	2.8
Birds (Poultry)	2.2	-	0.2	-	2

Source: *Statistical Abstracts of Punjab and Haryana* for Various Years.

The number of buffaloes as well as cows was more in Punjab than Haryana. But in both states more than 25 per cent of the buffaloes were not in milk. Similarly, more than fifty per cent of the cows were not in milk. It seems in Punjab, the size of poultry was large at the time of the early green revolution period. During the early period of the green revolution in Punjab there were 2.2 lakhs of poultry birds, whereas in Haryana the number was just twenty thousand.

To conclude, the level of agricultural development in terms of net sown area, level of irrigation and number of tubewells was higher in Punjab than Haryana. The institutional infrastructure like cooperative credit and regulated agricultural markets was also advanced in Punjab. The area, production and yield of food grains and non-food grains crops is also higher in Punjab than Haryana.

REFERENCES

1. Calculated from Ashok Mitra, *India's Population: Aspects of Quality and Control*, 1978.
2. In Punjab out of the total male agricultural labourers the proportion of Scheduled Castes was 70 per cent during 1961. For details, see Varinder Sharma: *Farm Workers of Punjab*, 2016, pp. 38-40.
3. The agrarian structure of the pre-green revolution period of the state of Punjab and PEPSU consisting of agricultural labourers, share wage servants and croppers for details see Daniel Thorner: *The Agrarian Prospects in India*, 1955, pp. 7-17.
4. For more details, see Simon Kuznet: *Modern Economic Growth*, 1966.
5. For more details, see Khem Singh Gill: *A Growing Agricultural Economy: Technological Changes, Constraints and Sustainability*, 1993.
6. Ibid.
7. For more details, see Varinder Sharma: *Agricultural Development and Character of Industrialisation: A Comparative Study of Punjab and Haryana*, M.Phil Dissertation (Unpublished), Panjab University, 1995.
8. The institutional loans comprised 19 per cent out of the total loans in Punjab and Haryana. In Bihar and Uttar Pradesh, it was just 3 to 5 per cent. See M.S. Randhawa: *Green Revolution*, 1962, pp. 81-88.

9. Ibid., p. 88.
10. Ibid.
11. For details, see R. Mohan, D. Jha and R.E. Evenson: "The Indian Agricultural Research System", *Economic and Political Weekly*, Vol. 8, No. 13, 1973, p. 23.
12. Minhas and Vaidyanathan: "Growth of Crop Output in India, 1951-54 to 1958-61: An Analysis by Component Elements", *Journal of Indian Society of Agricultural Statistics*, 1965.
13. For details, see G.S. Bhalla and G.K. Chadha: *Structure of Institutional Setup of Rural Punjab in the Year 2000*, FAO (Mimeo), 1980, p. 138.
14. In districts like Hissar many types of breeds of bulls, cows and buffaloes were popular. Income from farm livestock was around Rs. 15 lakhs. See Malcolm Lyall Darling: *Wisdom and Waste: In the Panjab Village*, 1934, p. 157.

CHAPTER 3

Agricultural Development in Punjab and Haryana in the Post Green Revolution Period

In the previous chapter, we have discussed the level of agricultural development in the early period of the green revolution and at the time of reorganisation of Punjab into two states, viz. Punjab and Haryana. In this chapter the present position of agricultural development in the two states is compared.

Change in Total Area in Punjab and Haryana

In two states, there was a very marginal change in the total area during the green revolution period (Table 3.1 and Figure 3.1). But it is interesting to note that the area under fallow lands in Punjab and Haryana continuously declined over these six decades. From the 1960s to 2000s, the fallow lands declined in Punjab by 167 thousand hectares and in Haryana by 139 thousand hectares. The reason behind this decline may be that fallow lands either came under cultivation or were used for various development works. It is interesting to note that the area under forests increased in Punjab from the 1960s onwards, but in Haryana it declined by 51 thousand hectares.

Table 3.1

Area in Punjab and Haryana (Average Over 2012-13 to 2014-15) ('000 Hectares)

Classification of Area	Punjab (i)	Percent-age	Haryana (ii)	Percent-age	Difference (i)-(ii)
Total Cropped Area	7858	-	6461	-	1397

Contd...

Cropping Intensity	190	-	184	-	6
Net Sown Area	4138	82.22	3511	80.32	627
Uncultivated Area	565	11.23	704	16.11	-139
Forests	259	5.15	39	0.89	220
Fallow Land	71	1.41	117	2.68	-46
Total Area	5033	100.00	4371	100.00	662

Source: *Statistical Abstracts of Punjab and Haryana* for Various Years.
Note: Percentages are out of the Total Area.

Figure 3.1

Classification of Area in Punjab and Haryana ('000 Hectares) During the Current Period

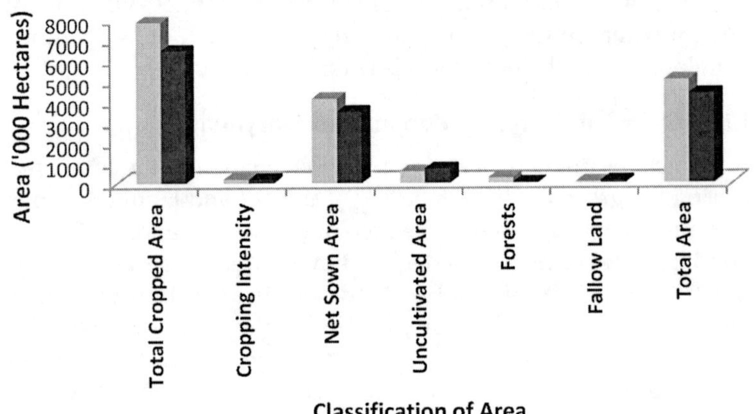

The uncultivated area continuously declined in both states during the last six decades. At present out of the total area the uncultivated area is 11.23 per cent in Punjab and in Haryana, it is 16.11 per cent. From the 1960s onwards the uncultivated area in Punjab reduced by 5 per cent and in Haryana it increased by 2 per cent. The growth rate of the total cropped area from the 1960s to the current period is given in Table 3.2.

Next we will compare the net sown area and cropping intensity.

Table 3.2
Annual Growth Rate of Total Cropped Area in Punjab and Haryana (Per Cent)

Time Period	Growth in Punjab State	Growth in Haryana State
1966-67 to 1980-81	2.00	1.00
1981-82 to 1995-96	0.09	0.50
1996-97 to 2014-15	(-)0.001	0.30
1966-67 to 2014-15	0.80	0.70

Source: *Statistical Abstracts of Punjab and Haryana* for Various Years.

In the early period of the green revolution, i.e. from 1966-67 to 1980-81, the total cropped area in Punjab increased by 2 per cent per annum and in Haryana by 1 per cent per year. In the next phase, i.e. from 1981-82 to 1995-96 the growth rate became lower relatively to the previous period, but in Punjab and Haryana; however, in the case of Haryana it was again less than Punjab. In the third phase of 1996-97 to 2014-15, the growth in area in case of Punjab turned into negative, and in case of Haryana, it was only 0.3 per cent. The overall growth from 1966-67 to 2014-15 in the total cropped area in Punjab and Haryana remained less than 1 per cent (Table 3.2) and results in semi-logarithmic models are given in Appendix-14 and time series data of total cropped area are given in Appendix-1.

In the percentage terms the total cropped area in Punjab increased by 52 per cent and in Haryana by 40 per cent. The overall total cropped area is more in Punjab.

Similarly, the net sown area in Punjab and Haryana increased, but not as rapidly as the total cropped area increased. In Punjab, it increased by 6 per cent from the early green revolution period to 2014-15 and in Haryana by 2 per cent. At present the net sown area in Punjab is 627 thousand hectares more than Haryana.

Lastly, the cropping intensity in both the states increased continuously from the early green revolution period to the current period. In Punjab it is 190 at present, was 133 during the 1960s. Similarly, for Haryana, it is 184 now and was 134 during the 60s (Table 3.1).

Thus, the expansion of the total cropped area, net sown area and the rise in cropping intensity became possible only due to advancement in irrigation and mechanisation. Before focusing on this, we will elaborate growth in the agriculture workforce and contribution of agriculture in the total state domestic product from the early green revolution period to the current period.

Change in Agriculture Workforce Structure

The new agriculture technology in Punjab completely changed the traditional agrarian structure[1]. With the green revolution a new class of casual agricultural labourers emerged in the agricultural developed regions. The demand for agricultural labourers increased tremendously in Punjab and Haryana. At the time of harvesting and transplantation of paddy there remains a big gap in supply and demand of labour. This gap is filled by the migrant agricultural labourers who come from Bihar and Uttar Pradesh[2].

The number of agriculture workers in Punjab and Haryana is given in Tables 3.3A and 3.3B. The total number of agricultural workers (cultivators and agriculture labourers) in Punjab is 35.22 lakhs, whereas in Haryana, the number is 40.08 lakhs.

Table 3.3A

Agriculture Workforce in Punjab and Haryana (2011) (Lakhs)

Categories	Punjab (i)	Percent-age	Haryana (ii)	Percent-age	Difference (i)-(ii)
Agricultural Labourers	15.88	45.09	15.28	38.12	0.6
Cultivators	19.34	54.91	24.8	61.88	-5.46
Agricultural Workers	35.22	100	40.08	100	-4.86

Source: *Census of India 2011.*

Out of the total agricultural workers in Punjab 45.09 per cent are agriculture labourers and 54.91 per cent are cultivators. In Haryana, the percentage of agriculture labourers is 38.12 per cent and of cultivators, 61.88 per cent. If we compare the growth from 1961 to 2011 then in Punjab, the number of agricultural labourers grew by 3.09 per cent per annum and cultivators by 0.38 per cent. The number of agricultural labourers in Haryana grew by 4.16 per cent per annum and of cultivators by 0.60 per cent per annum.

Table 3.3B

Change in Agriculture Workforce Structure: 1961-2011

Type of Agriculture Workers	Punjab (Increase '+')	Growth Rate	Haryana (Increase '+')	Growth Rate
Agriculture Labourers	+12.54	3.09	+13.29	4.16
Cultivators	+3.32	0.38	+6.42	0.60
Agriculture Workers	+15.86	1.2	+19.71	1.36

Source: *Census of Punjab and Haryana for 1961 and 2011.*

Overall agriculture workers made a growth of 1.2 per cent per annum in Punjab. On the other hand in Haryana, the growth in the number of agricultural labourers is almost near to Punjab but number of cultivators increased more rapidly in Haryana than Punjab. It means the division of land among cultivator households in Haryana is faster than Punjab. In Punjab the number of small landholdings decreased over the periods and even the average size of landholdings increased in Punjab[3].

Deceleration in the Share of the Primary Sector in Net State Domestic Product (NSDP)

The Fisher-Clark hypothesis maintains that with economic development the share of the primary sector in income and employment declines and of the secondary and tertiary sectors increases. In the previous chapter, in Table 2.3, we noted that the share of the primary sector was more than fifty per cent in both states in the mid-1960s. Over the years, the share of primary as well as of agriculture and animal husbandry in total net state domestic product sharply declined in both the states. At present, the share of agriculture and animal husbandry in net state domestic is 29 per cent in Punjab and 20 per cent in Haryana (Table 3.4).

Table 3.4

Percentage Share of the Primary Sector in Net State Domestic Product (At 2011-12 Prices) 2014-15

Industry	Punjab (i)	Haryana (ii)	Difference (i)-(ii)
Agriculture and Animal Husbandry	29.28	19.95	9.33
Forestry	2.63	1.19	1.44

Contd...

Fishery	0.23	0.27	-0.04
Total Primary Sector	29.3	20.02	9.28

Source: *Statistical Abstracts of Punjab and Haryana* for Various Years.

The overall share of the total primary sector in net state domestic product has come down to 29 per cent in Punjab and 20 per cent in Haryana. It means from the 1960s onwards the secondary and tertiary sector developed faster in these two states[4]. The annual compound growth rate of agriculture in total NSDP is given below.

Table 3.5

Compound Annual Growth Rate of Net State Domestic Product (Agriculture) of Punjab and Haryana

Years	States	
	Punjab	Haryana
1970-71 to 1980-81	2.72	2.04
1981-82 to 1992-93	11.69	13.45
1993-94 to 2011-12	4.72	6.39
2012-13 to 2014-15	0.07	0.77

Source: *Statistical Abstracts of Punjab and Haryana* for Various Years.

The overall share of agriculture in total NSDP from 1966-67 onwards is given in Appendix-28.

Changes in Net Irrigated Area and Development in Different Sources of Irrigation

As we have already discussed in the previous chapter that from the colonial period to the early green revolution period the governments mainly focused on the canal-based irrigation in Punjab and Haryana. After 1966, the governments adopted a policy to develop tubewell irrigation in the states. In the beginning, the tubewells played a minor role in irrigation in the two states. The actual boom in tubewell irrigation emerged after the land consolidation, rise in the institutional credit, and fast rural electrification of villages in Punjab and Haryana[5].

This development in irrigation in the two states is described in Tables 3.6 and 3.7. In both states the net irrigated area almost doubled in Punjab and Haryana compared to the early period of the green revolution.

Table 3.6

Net Irrigated Area by Different Sources in Punjab and Haryana (Average Over 2012-13 to 2014-15) ('000 Hectares)

Level of Irrigation	Punjab (i)	Haryana (ii)	Difference (i)-(ii)
Net Irrigated Area	4125	3002	1123
Area Irrigated by Canals	1156	1235	-79
Area Irrigated by Tubewells	2969	1765	1204
Percentage of Net Irrigated Area to Net Sown Area (%)	99.6	85.5	14.1
Percentage of Gross Irrigated to Gross Cropped Area (%)	98.53	89.78	8.75

Source: *Statistical Abstracts of Punjab and Haryana* for Various Years.

Table 3.7

Net Irrigated Area by Different Sources in Punjab and Haryana (Average Over 2012-13 to 2014-15) ('000 Hectares)

Sources of Irrigation	Punjab	Percent-age	Haryana	Percent-age	Difference
Canals	1156	28.02	1235	41.14	-79
Tubewells	2969	71.98	1767	58.86	1202
Other Sources*	-	-	-	-	-
Total Irrigated Area	4125	100.00	3002	100.00	1123

Source: *Statistical Abstract of Punjab and Haryana* for Various Years.
Note: * Less Than 500 Hectares.

Similarly, the tremendous changes occurred in the area under tubewell irrigation in Punjab as well as in Haryana. At present out of the total irrigated area, 71.98 per cent of the area is irrigated by tubewells in Punjab but in Haryana, it is just 58.86 per cent.

Figure 3.2A

Percentage of Net Irrigated Area by Different Sources in Punjab During the Current Period

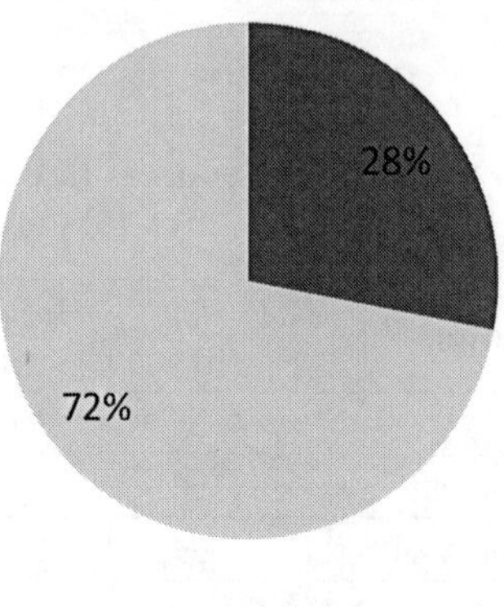

Figure 3.2B

Percentage of Net Irrigated Area by Different Sources in Haryana During the Current Period

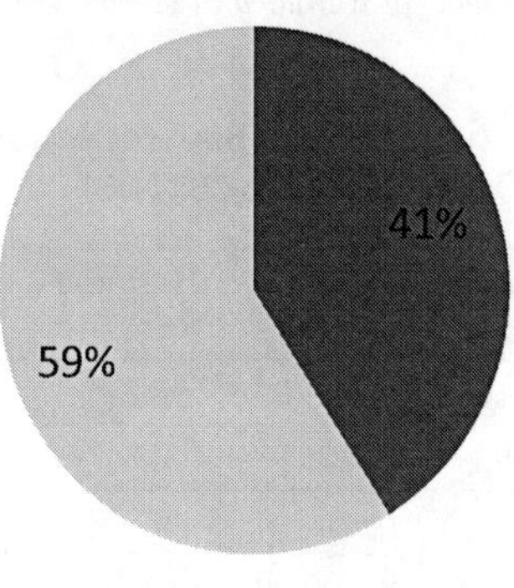

The area under canal irrigation is just 28 per cent in Punjab and 41 per cent in Haryana. The percentage of net and gross irrigated area is around 99 per cent in Punjab and more than 80 per cent in Haryana (Table 3.6).

Use of Agriculture Inputs in Punjab and Haryana

In addition to the advancement in irrigation, the other agricultural inputs which were essential for the successful cultivation of

HYV seeds was fertilisers. The use of fertilisers at present is 1787 thousand tonnes in Punjab and 1274 thousand tonnes in Haryana. From the early green revolution period, the use increased by eight times in Punjab and almost fifteen times in Haryana. The per hectare use of fertiliser in Punjab is 227 kgs and in Haryana, it is 197 kgs. In both the states, it increased four times from the earlier periods of the 1960s (Table 3.8).

The number of tubewells in both the states increased sharply from the earlier period of the green revolution. At present in Punjab, there are 13.98 lakh tubewells and in Haryana, 8 lakhs.

Table 3.8

Main Agriculture Inputs in Punjab and Haryana (Average Over 2012-13 to 2014-15)

Inputs	Punjab (i)	Haryana (ii)	Difference (i)-(ii)
Fertilisers (000 Tonnes)	1787	1274	513
Tubewells used for Irrigation (Lakhs)	13.98	8	5.98
Consumption of Electricity (Million Kwh)	2074	995	1079
Number of Tractors (Lakhs)	5.17	5.55	
Loans Advanced by Primary Agriculture Credit Societies (Rs. Lakhs)	806181	752595	53586
Number of Agriculture Markets	152	107	45

Source: *Statistical Abstracts of Punjab and Haryana* for Various Years.

The increasing number of tubewells also posed a serious problem of the declining water table in both states. The over exploitation of ground water is also a cause of concern in both the states[6].

The electricity in agriculture is provided free by the state to the farmers in Punjab[7]. The use of electricity is also quite high in agriculture in Punjab than Haryana. The use of electricity in Haryana is over 995 million kwh but in Punjab this figure touches 2074 million kwh.

Since the beginning of the green revolution most of the farm operations are fully mechanised. The tractor is the main farm machinery which was introduced at the beginning of the green revolution in two states. Many scholars studied the impact of tractors on the employment of casual agricultural and annual farm servants[8]. The number of tractors in Punjab and Haryana mushroomed over the decades. At present there are around five lakh tractors in each state (Table 3.8). Further, it is also accepted that with this mechanisation of the farm employment avenues will increase. According to S.S. Johl, who remained in favour of this model: "Mechanisation of agriculture will create new avenues of jobs such as those of drivers, mechanics and semi-skilled labourers attending to machinery and machine operations"[9].

To make the use of agricultural inputs more effective, the state governments established the Cooperative Credit Societies in each village. For the long term agriculture credit, there are State Land Mortgage Banks and Commercial Banks. From the earlier green revolution period to the present time the agricultural loans of Cooperative Credit Societies increased many times in both states. At present, Rs.8061.81 crore loans are advanced in Punjab and Rs.7525.95 crores in Haryana (table 3.8).

Growth in Area Under Foodgrain Crops in Punjab and Haryana in the Post Green Revolution Period

In this section and in the next sub-sections, we have studied the expansion in area, production and yield of main foodgrain crops in Punjab and Haryana. From the earlier period of the green revolution to the present time, the area under foodgrains in both states increased by more than one and a half times. It is interesting to note that in both states there are now only two dominant crops, i.e. wheat and paddy. The area under these two crops from 1966-67 onwards is given in Appendix-1.

Figure 3.3A

Percentage of Area Under Main Foodgrain Crops in Punjab During the Current Period

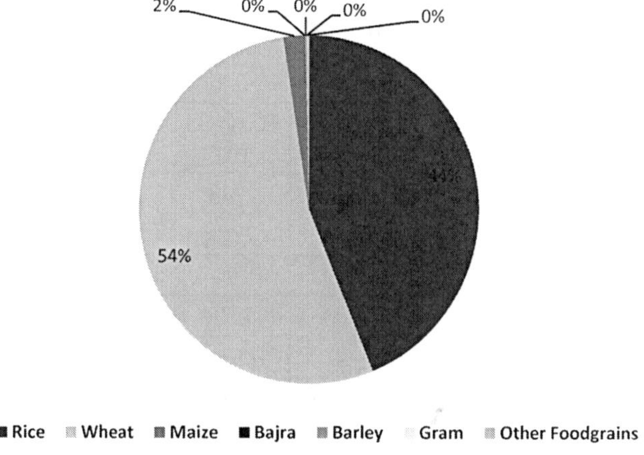

Figure 3.3B

Percentage of Area Under Main Foodgrain Crops in Punjab During the Current Period

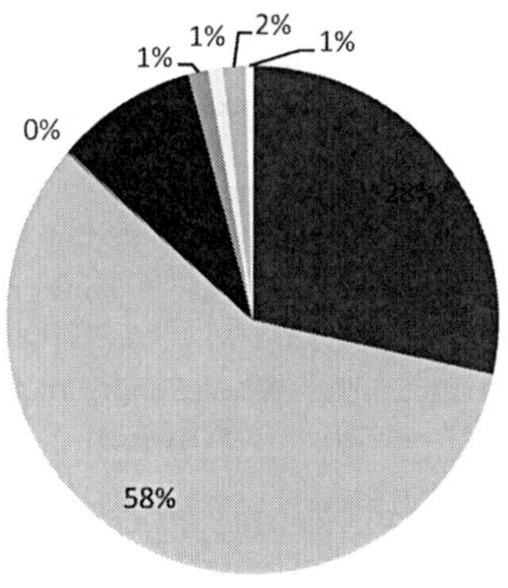

The area under other crops has declined over the green revolution period. In Punjab more than 50 per cent of the area is under the wheat crop and 40 per cent under rice. On the other hand, in Haryana more than 50 per cent of the area is under wheat and 28 per cent under rice. The state of Punjab is far ahead than Haryana in terms of cultivation of the rice crop (Table 3.9).

Table 3.9

Area Under Main Foodgrain Crops in Punjab and Haryana (Average Over 2012-13 to 2014-15) ('000 Hectares)

Crops	Punjab (i)	Percent-age	Haryana (ii)	Percent-age	Difference (i)-(ii)
Rice	2864	43.83	1243	28.37	1621
Wheat	3510	53.72	2541	57.99	969
Maize	129	1.97	9	0.21	120
Bajra	2	0.03	403	9.2	-401
Jowar	@	@	57	1.3	-
Barley	12	0.18	41	0.94	-29
Gram	2	0.03	65	1.48	-63
Other Foodgrains	15	0.23	23	0.52	-8
Total Area Under Foodgrains	6534	100	4382	100	2152

Source: *Statistical Abstracts of Punjab and Haryana* for Various Years.
Note: (i) @ Area under jowar is below 500 hectares.
(ii) Percentages are out of the total area.

The growth rates are calculated by the semi-logarithmic linear trend models and detailed results are given in Appendices 16 to 19. The growth in area under two principal crops, i.e. wheat and paddy are given in (Tables 3.10 and 3.11). From 1966-67 to 1980-81, in Punjab the area under wheat increased by 3 per cent per annum and rice by 10 per cent per annum.

Table 3.10

Growth in Area Under Wheat and Rice in Punjab

Years	Wheat (Per Cent)	Rice (Per Cent)
1966-67 to 1980-81	3.00	10.00
1981-82 to 1995-96	0.07	4.00
1996-97 to 2014-15	0.04	1.00
1966-67 to 2014-15	1.00	5.00

Source: *Statistical Abstracts of Punjab.*

During the same time period in Haryana, area under wheat increased by 4 per cent annually and under rice by 6 per cent.

Table 3.11

Growth in Area Under Wheat and Rice in Haryana

Years	Wheat (Per Cent)	Rice (Per Cent)
1966-67 to 1980-81	4.00	6.00
1981-82 to 1995-96	1.00	3.00
1996-97 to 2014-15	1.00	1.00
1966-67 to 2014-15	2.00	4.00

Source: *Statistical Abstracts of Haryana.*

In the next period, i.e. from 1981-82 to 1995-96, in the Punjab and Haryana area under wheat increased by 0.07 and 1.00 per cent per annum respectively. On other hand, the area under rice increased faster by 3 to 4 per cent in Haryana and Punjab. It means during this phase other foodgrain crops were replaced by rice faster. In current period, i.e. from 1996-97 to 2014-15, the area under wheat expanded by less than one per cent in Punjab and in Haryana by one per cent. In both the states the area under rice increased by one per cent per annum. It means the maximum area under wheat increased in both states during 1966-67 to 1980-81 and under rice during 1996-97 to 2014-15. Overall, from 1966-67 to 2014-15, the area under rice increased more rapidly in both states than wheat. Some economists also not in favour of this established cropping pattern of two crops suggest the diversification of crops[10]. But farmers are not willing to shift towards other crops due to an assured market of wheat and paddy.

Production of Main Foodgrain Crops

In this section, we have compared the production of foodgrains over the years in Punjab and Haryana. The production of total foodgrains increased by 6 per cent per annum in Punjab and by 4 per cent in Haryana during the early 1960s to 1980s. In case of Punjab from 1981-82 to 1995-96, the growth in production, became almost half compared to the previous of foodgrain period and in Haryana remained the same.

From the 1960s onwards to the present time, the production of rice increased by thirty times in Punjab and by sixteen times in Haryana.

Table 3.12

Production of Main Foodgrain Crops in Punjab and Haryana (Average Over 2012-13 to 2014-15) ('000 Metric Tonnes)

Crops	Punjab (i)	Percent-age	Haryana (ii)	Percent-age	Difference (i)-(ii)
Rice	11253	39.86	3996	24.51	7257
Wheat	16437	58.22	11208	68.74	5229
Maize	484	1.71	23	0.14	461
Bajra	2	0.01	763	4.68	-761
Jowar	@	@	29	0.18	-
Barley	45	0.16	141	0.86	-96
Gram	2	0.01	56	0.34	-54
Other Foodgrains	11	0.04	88	0.54	-77
Total Production	28234	100	16304	100	11930

Source: *Statistical Abstracts of Punjab and Haryana* for Various Years.
Note: (i) @ Production of jowar is below 500 metric tonnes.
(ii) Percentages are out of total production.

Similarly, the production of wheat increased by six times in Punjab and ten times in Haryana. From this we may conclude that in both states production of rice increased faster than any foodgrain crops except barley in Haryana which continuously declined in both states. Next, in Table 3.13 the growth in production during different time periods is shown. The time series data for total foodgrains and for main foodgrain crops, i.e. wheat and rice is given in Appendices 2 and 6 along with the semi-logarithmic growth models in Appendices 14 and 15.

Table 3.13

Growth in Production Under Total Foodgrains in Punjab and Haryana

Years	Punjab	Haryana
1966-67 to 1980-81	6.00	4.00
1981-82 to 1995-96	3.00	4.00
1996-97 to 2014-15	1.00	2.00
1966-67 to 2014-15	3.00	4.00

Source: *Statistical Abstracts of Punjab and Haryana.*

Further, during 1996-97 to 2014-15 in Punjab the growth remained around 1 per cent and in Haryana around 2 per cent. From 1966-67 to 2014-15, the growth in production of foodgrains remained around 3 per cent and in Haryana around 4 per cent.

Figure 3.4

Production of Main Foodgrain Crops in Punjab and Haryana ('000 Metric Tonnes) During the Current Period

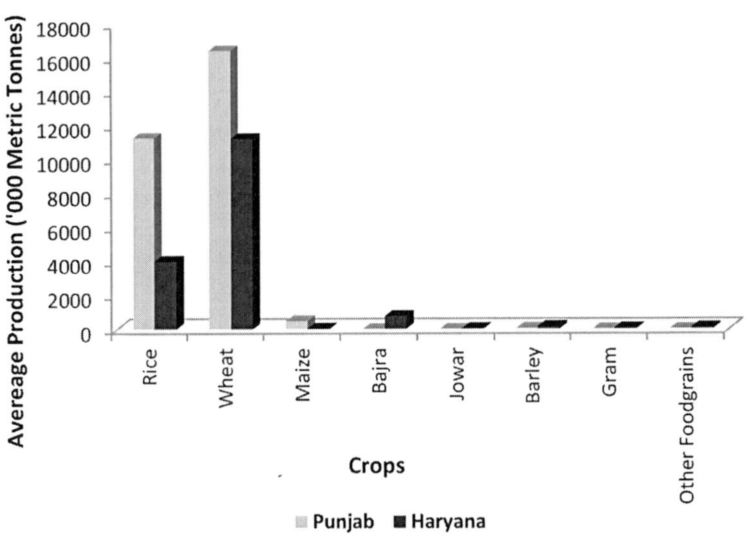

During the early green revolution period, i.e. from 1966-67 to 1980-81, the production of wheat increased by 6 per cent annually in Punjab and by 5 per cent in Haryana. It is interesting to note in this period, the production of rice increased by 17 per cent in Punjab and by 12 per cent in Haryana.

Table 3.14

Growth in Production of Wheat and Rice in Punjab

Years	Wheat (Per Cent)	Rice (Per Cent)
1966-67 to 1980-81	6.00	17.00
1981-82 to 1995-96	3.00	5.00
1996-97 to 2014-15	1.00	2.00
1966-67 to 2014-15	3.00	7.00

Source: *Statistical Abstracts of Punjab and Haryana.*

The second phase lies between 1981-82 to 1995-96. The growth in production of wheat remained 3 per cent, which is half the growth of the previous period. The growth in Haryana remained the same as in the previous period.

Table 3.15

Growth in Production of Wheat and Rice in Haryana

Years	Wheat (Per Cent)	Rice (Per Cent)
1966-67 to 1980-81	5.00	12.00
1981-82 to 1995-96	5.00	4.00
1996-97 to 2014-15	2.00	3.00
1966-67 to 2014-15	5.00	5.00

Source: *Statistical Abstracts of Punjab and Haryana.*

In the third phase, i.e. from 1996-97 to 2014-15, the growth rate in production of wheat further reduced by 2 per cent in Punjab and 3 per cent in Haryana. Similarly, the production of rice reduced by 3 per cent in Punjab and by one per cent in Haryana.

From 1966-67 to 2014-15, the production of wheat increased by 3 per cent per annum in Punjab and in Haryana by 5 per cent. In contrary to this, the growth in production of rice in Punjab remained around 7 per cent and in Haryana remained 5 per cent. Overall growth in production of foodgrains remained highest in Punjab and Haryana during the early period of the green revolution, i.e. from 1966-67 to 1980-81. After this upto 1995-96, it either declined or stagnated. But during this period, i.e. from 1996-97 to 2014-15 it again started increasing.

Change in Yield of Foodgrains in Punjab and Haryana

The yield of two principal crops, i.e. wheat and paddy increased by three times in two states from the early green revolution period to the current period. At present except for the bajra crop, the yield of other crops is higher in Punjab than Haryana.

Table 3.16
Yield of Main Foodgrain Crops in Punjab and Haryana (Average Over 2012-13 to 2014-15) (Per Hectare Kg)

Crops	Punjab (i)	Haryana (ii)	Difference (i)-(ii)
Rice	3929	3210	719
Wheat	4682	4385	297
Maize	3743	2617	1126
Bajra	935	1910	-975
Jowar	@	508	-
Barley	3760	3511	249
Gram	914	880	34

Source: *Statistical Abstracts of Punjab and Haryana* for Various Years.
Note: @ Yield cannot be calculated owing to low levels of production and area under cultivation.

Figure 3.5
Yield of Main Foodgrain Crops in Punjab and Haryana (Kg/Hectare) During the Current Period

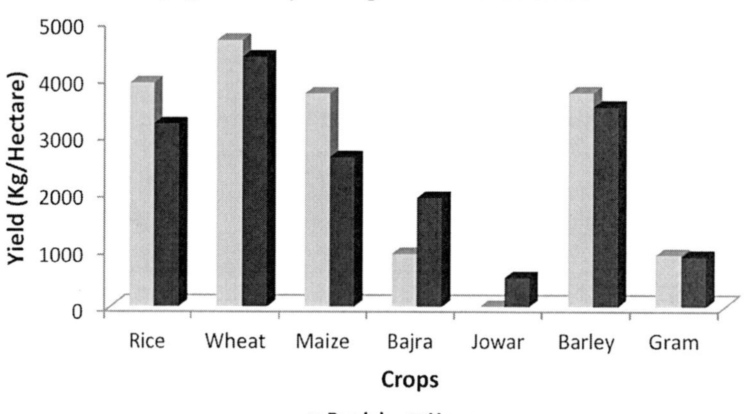

The growth in yield of two principal crops, i.e. wheat and rice is given in Tables 3.17 and 3.18. In the early period of the green revolution, i.e. from 1966-67 to 1980-81, the growth in the yield of wheat remained 3 per cent per annum in Punjab, but in Haryana, it remained less than one per cent. The semi-logarithmic growth models of yield of rice and wheat from 1966-67 onwards in Punjab and Haryana are given in Appendices 16 and 19.

Table 3.17

Growth in Yield of Rice and Wheat in Punjab

Years	Wheat (Per Cent)	Rice (Per Cent)
1966-67 to 1980-81	3.00	6.00
1981-82 to 1995-96	2.00	0.07
1996-97 to 2014-15	0.60	1.00
1966-67 to 2014-15	1.00	2.00

Source: *Statistical Abstracts of Punjab.*

From the period of 1981-82 to 1995-96, the growth in yield of wheat in Haryana remained around 3 per cent but in Punjab remained 2 per cent. During the third phase, i.e. from 1996-97 to 2014-15, the growth rate in yield of this crop remained less in Punjab than Haryana. From 1996-97 to 2014-15, the growth rate in Punjab and Haryana further declined but decline was faster in Punjab than Haryana. Overall from 1966-67 to 2014-15, the growth in yield of wheat remained around 1 per cent in Punjab and 2 per cent in Haryana.

Table 3.18

Growth in Yield of Rice and Wheat in Haryana

Years	Wheat (Per Cent)	Rice (Per Cent)
1966-67 to 1980-81	0.05	5.00
1981-82 to 1995-96	3.00	(-)0.03
1996-97 to 2014-15	1.00	1.00
1966-67 to 2014-15	2.00	1.00

Source: *Statistical Abstracts of Haryana.*

Now the yield of rice increased by 6 per cent per annum in Punjab and in Haryana by 5 per cent. In the next period, i.e. from 1981-82 to 1995-96, the growth rate of rice slowed down and it turned into less than one per cent in Punjab and in Haryana it became negative. From 1996-97 to 2014-15, the yield level growth rates are almost the same in Punjab and Haryana. During these years, the growth rate of yield remained around 1 per cent.

Overall from 1966-67 to 2014-15, the growth rate of yield of rice was 2 per cent in Punjab, and 1 per cent in Haryana.

To conclude, in the early periods of the green revolution, i.e. from 1966-67 to 1980-81, the growth in area, production and yield of two principal crops i.e. wheat and paddy and other food grains remained higher in two states relatively to the post green revolution periods.

Area Under Non-Foodgrains in Punjab and Haryana

In the preceding section, we have discussed the area, production and yield of foodgrains in Punjab and Haryana. Here in this section, we have compared how the green revolution impacted the growth in area under non-foodgrain crops in these agriculturally developed states. In absolute terms the area under non-foodgrain crops declined in Punjab from the early green revolution period to the current period by almost 400 thousand hectares.

Table 3.19

Area under Major Non-Foodgrain Crops in Punjab and Haryana (Average Over 2012-13 to 2014-15) ('000 Hectares)

Crops	Punjab (i)	Percent-age	Haryana (ii)	Percent-age	Difference (i)-(ii)
Rape and Mustard	29	4.96	526	42.45	-497
Sugarcane	89	15.21	99	7.99	-10
Cotton	449	76.75	603	48.67	-154
Other Non-Foodgrains	18	3.08	11	0.89	7
Total Area Under Non-Foodgrains	585	100	1239	100	-654

Source: *Statistical Abstracts of Punjab and Haryana* for Various Years.

But in Haryana, it increased by 1400 thousand hectares. In Punjab except for cotton, the area under rape and mustard and sugarcane declined but in Haryana the area under all these crops increased from the 1960s onwards to this period. The area under non-foodgrains from 1966-67 onwards in Punjab and Haryana are given in Appendices 9 and 11.

Figure 3.6A

Percentage of Area Under Main Non-Foodgrain Crops in Punjab During the Current Period

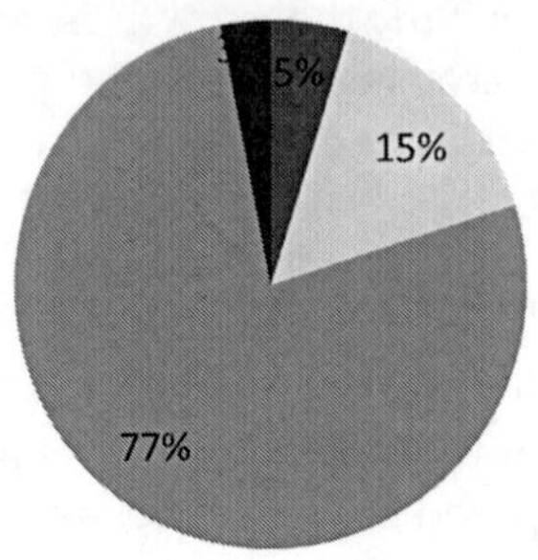

Figure 3.6B

Percentage of Area Under Main Non-Foodgrain Crops in Haryana During the Current Period

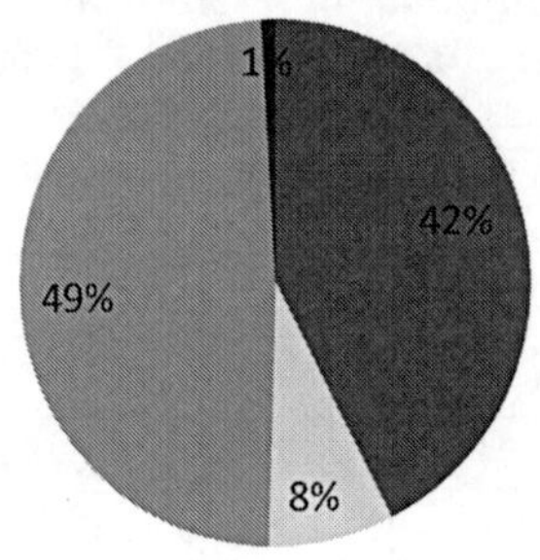

Table 3.20

Growth in Area Under Total Non-Foodgrain Crops in Punjab and Haryana

Years	Punjab (Per Cent)	Haryana (Per Cent)
1966-67 to 1980-81	0.07	4.00
1981-82 to 1995-96	0.02	2.00
1996-97 to 2014-15	(-)2.00	(-)0.06
1966-67 to 2014-15	(-)0.08	2.00

Source: *Statistical Abstracts of Punjab and Haryana.*

The semi-logarithmic growth models of area in Punjab and Haryana are given in Appendices 20 and 24. The growth in area under total non-foodgrains are given in Table 3.20.

The annual growth rate of the area of non-foodgrain crops is given in Table 3.20. During the early green revolution period i.e. from 1966-67 to 1980-81, the area under non-foodgrain crops in Punjab increased merely by 0.07 per cent per annum, whereas in Haryana it increased by 4 per cent per annum. During the 1980s and early 1990s, i.e. from 1981-82 to 1995-96, it again declined to 0.05 per cent in Punjab compared to the previous period, and in Haryana the growth rate also became half relative to the early green revolution period but remained more than Punjab. The period of the 1990s, i.e. from 1996-97 to 2014-15, the growth in area in both states turned into negative in Punjab and the area under non-foodgrains declined more sharply than Haryana. Overall, from 1966-67 to 2014-15, in the Punjab area under non-foodgrains declined by (-)0.08 per cent and (-) 0.06 per cent in Haryana. In both states it declined during different time periods but in Punjab it declined more sharply than Haryana.

The decline in area under non-foodgrain crops is primarily due to two states adopting crops mainly, i.e. wheat and paddy which are more remunerative than the non-foodgrain crops. Moreover, the prices of non-foodgrain crops are highly volatile and farmers do not want to take risks by growing non-foodgrain crops. On the other hand, the price and market for wheat and paddy is assured with minimum support price and purchased by the government agencies. But the farmers demand minimum support prices as per the Swaminathan Committee recommendations.

Table 3.21

Growth in Area Under Major Non-Foodgrain Crops in Punjab

Years	Sugarcane (Per Cent)	Cotton (Per Cent)	Oilseeds (Per Cent)
1966-67 to 1980-81	(-)4.00	3.00	(-)2.00
1981-82 to 1995-96	0.08	0.03	(-)0.07
1996-97 to 2014-15	(-)3.00	(-)1.00	(-)7.00
1966-67 to 2014-15	0.09	0.07	(-)4.00

Source: *Statistical Abstracts of Punjab.*

The area under main non-foodgrain crops, i.e. sugarcane, cotton and oil seeds remained fluctuating in Punjab and Haryana over the years (Tables 3.21 and 3.22). In the early green revolution period, i.e. from 1966-67 to 1980-81, the area under sugarcane declined by (-)4 per cent annually and in Haryana increased by 0.03 per cent per annum. It is interesting in the next phase during the 1980s, it increased by 0.008 per cent in Punjab but in Haryana declined by 0.02 per cent annually.

Table 3.22

Growth in Area Under Major Non-Foodgrain Crops in Haryana

Years	Sugarcane (Per Cent)	Cotton (Per Cent)	Oilseeds (Per Cent)
1966-67 to 1980-81	0.03	5.00	1.00
1981-82 to 1995-96	(-)0.02	(-)0.008	7.00
1996-97 to 2014-15	(-)3.00	(-)0.04	(-)3.00
1966-67 to 2014-15	(-)7.00	2.00	3.00

Source: *Statistical Abstracts of Haryana.*

Further, in the period of the 1990s in Punjab and Haryana, the area under sugarcane declined by 3 per cent per annum. Overall from 1966-67 to 2014-15, the area under sugarcane in Punjab increased by 0.09 per cent annually but in Haryana declined by 7 per cent per annum. The main reason behind this decline in area under sugarcane is due to delay in payments to farmers from the sugar mill owners.

The area under cotton during the 1960s increased by 3 per cent per annum in Punjab and by 5 per cent in Haryana. During the 1980s, in Punjab the area expansion under cotton in Punjab slowed down compared to the 1960s but remained positive. The growth in this period was 0.03 per cent but in Haryana, it turned into negative. The period of the 1990s, shows the growth in the area under cotton in both states became negative but in Punjab it reduced more sharply than Haryana. The reason may be owing to frequent failures of the cotton crop due to the attack of whiteflies and water logging problems in the cotton growing area during this period, especially in Punjab. The continuous failure of cotton crops also de-established the economy of farmers and many of them started committing suicide to end these vagaries in life[11].

The rapid fluctuations in the prices of non-foodgrains are also the cause of the declining area under non-foodgrains[12].

Overall from 1996-67 to 2014-15, the area under cotton in Punjab and Haryana increased marginally in Punjab by 0.07 per cent per annum and in Haryana by 2 per cent per annum. The area under oilseeds in Punjab decreased by 2 per cent per annum but in Haryana increased by 1 per cent annually during the 1960s. From 1981-82 to 1995-96, the area under oilseeds declined by 0.07 per cent per annum in Punjab, but increased by 7 per cent per annum in Haryana. From 1996-97 to 2014-15, it declined in both states. Overall from 1966-67 to 2014-15, the area under oilseeds in Punjab declined by 4 per cent annually but in Haryana increased by 3 per cent per annum.

Production of Non-Foodgrain Crops in Punjab and Haryana

Here, in this section we have tried to compare the production of non-foodgrain crops in Punjab and Haryana. Compared to the periods of the 1960s in the current period the production of non-foodgrain crops increased in Haryana, but decreased in Punjab (Table 3.23). The production of non-foodgrains from 1966-67 onwards in Punjab and Haryana are given in Appendices 10 and 12.

Table 3.23

Production of Major Non-Foodgrain Crops in Punjab and Haryana (Average Over 2012-13 to 2014-15) ('000 Tonnes)

Crops	Punjab (i)	Percent-age	Haryana (ii)	Percent-age	Difference (i)-(ii)
Rape and Mustard	37	4.31	848	43	-811
Sugarcane	545	63.45	742	37.63	-197
Cotton	253	29.45	360	18.26	-107
Other Non-Foodgrains	24	2.79	22	1.12	2
Total Production of Non-Foodgrains	859	100.00	1972	100.00	-1113

Source: *Statistical Abstracts of Punjab and Haryana* for Various Years.

Figure 3.7

Production of Main Non-Foodgrain Crops in Punjab and Haryana ('000 Tonnes) During the Current Period

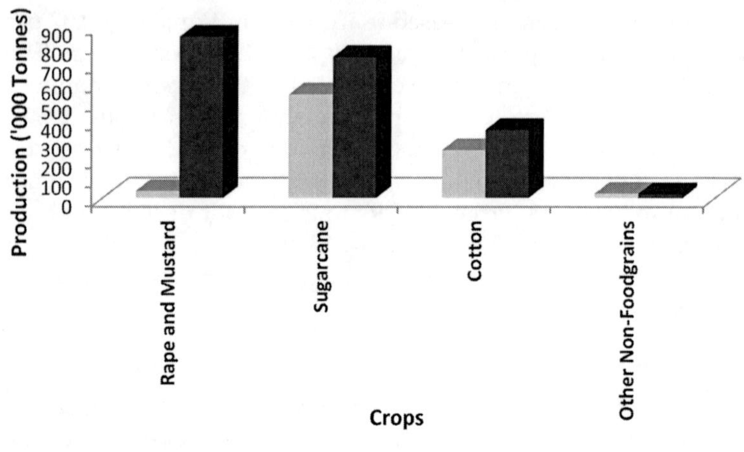

The annual compound growth rates given in Table 3.24 highlight the production of non-foodgrains in Punjab during the early 1960s increased by 9 per cent annually, but in Haryana by merely 6 per cent (Table 3.25). The semi-logarithmic models of growth of total and major non-foodgrain crops in Punjab and Haryana are given in Appendices 20 and 27.

Table 3.24

Growth in Production of Major Non-Foodgrain Crops in Punjab

Years	Total Non-Foodgrain Production	Sugarcane (Per Cent)	Cotton (Per Cent)	Oilseeds (Per Cent)
1966-67 to 1980-81	9.00	0.01	3.00	(-)2.00
1981-82 to 1995-96	2.00	0.09	4.00	3.00
1996-97 to 2014-15	(-)1.00	(-)3.00	3.00	(-)5.00
1966-67 to 2014-15	4.00	0.02	1.00	(-)20.00

Source: *Statistical Abstracts of Punjab.*

From 1981-92 to 1995-96, the production of non-foodgrains reduced in Punjab and Haryana by 7 per cent in Punjab and by 2 per cent in Haryana compared to the previous period of the 1960s. Overall, from 1996-97 to 2014-15, in Punjab and Haryana production of non-foodgrain crops reduced in both states but it reduced faster in Punjab, i.e. by 1 per cent per annum and in Haryana by 0.006 per cent.

Table 3.25

Growth in Production of Major Non-Foodgrain Crops in Haryana

Years	Total Non-Foodgrain Production	Sugarcane (Per Cent)	Cotton (Per Cent)	Oilseeds (Per Cent)
1966-67 to 1980-81	6.00	-0.01	5.00	4.00
1981-82 to 1995-96	4.00	0.02	6.00	11.00
1996-97 to 2014-15	-0.006	-1.00	4.00	2.00
1966-67 to 2014-15	6.00	0.06	4.00	6.00

Source: *Statistical Abstracts of Haryana.*

At the crop level, the growth in production of sugarcane in both states from 1966-67 to 1980-81 is contrary to each other. In Punjab, it increased by 0.01 per cent but in Haryana declined by 0.01 per cent. From 1981-82 to 1995-96, the production of sugarcane increased by 0.09 per cent in Punjab and in Haryana by 0.02 per cent. During the 1990s, the production declined more sharply in Punjab by 3 per cent and in Haryana by 1 per cent. Overall from 1966-67 to 2014-15, the growth in production of sugarcane remained less than one per cent.

It is interesting to note that in both the states the growth in production of cotton remained positive throughout all the years. During the 1980s in both states the growth in production remained high compared to other periods. Overall growth in production from 1966-67 to 2014-15, the growth remained 1 per cent in Punjab which is less by three per cent in Haryana.

The growth in production of oilseeds remained negative in Punjab during all the periods except during the 1980s. In Haryana, the maximum growth remained during the 1980s in all the other periods; it is positive and more than Punjab.

Yield of Main Non-Foodgrains in Punjab and Haryana

The growth in the yield of non-foodgrain crops is shown in Table 3.26. The growth in yield level of sugarcane remained highest in Punjab during 1966-67 to 1980-81. After this it sharply declined during the 1980s period and slightly increased by the 1990s. Overall yield of major non-foodgrain crops in Punjab and Haryana are given in Appendices 10 and 12.

Table 3.26

Yield of Major Non-Foodgrain Crops in Punjab and Haryana (Average Over 2012-13 to 2014-15) (Kgs Per Hectare)

Crops	Punjab (i)	Haryana (ii)	Difference (i)-(ii)
Rape and Mustard	1282	1598	-316
Sugarcane	6090	7456	-1366
Cotton	523	627	-104

Source: *Statistical Abstracts of Punjab and Haryana* for Various Years.

Overall growth in yield of sugarcane remained less than one per cent from 1966-67 to 2014-15. The semi-logarithmic growth models of yield of major crops are given in Appendices 21 to 27.

Table 3.27

Growth in Yield of Major Non-Foodgrain Crops in Punjab

Years	Sugarcane (Per Cent)	Cotton (Per Cent)	Oilseeds (Per Cent)
1966-67 to 1980-81	4.00	(-)0.01	0.005
1981-82 to 1995-96	0.01	4.00	3.00
1996-97 to 2014-15	0.03	5.00	1.00
1966-67 to 2014-15	0.08	0.02	1.00

Source: *Statistical Abstracts of Punjab.*

In Haryana, it remained negative during the 1960s. But growth in yield of sugarcane remained around 2 per cent during the 1980s and 1990s. Overall growth remained around 1 per cent from 1966-67 to 2014-15. It means in both states sugarcane was not cultivated at large (Tables 3.27 and 3.28).

Table 3.28
Growth in Yield of Major Non-Foodgrain Crops in Haryana

Years	Sugarcane (Per Cent)	Cotton (Per Cent)	Oilseeds (Per Cent)
1966-67 to 1980-81	(-)0.04	(-)1.00	3.00
1981-82 to 1995-96	2.00	6.00	4.00
1996-97 to 2014-15	2.00	5.00	2.00
1966-67 to 2014-15	1.00	2.00	2.00

Source: *Statistical Abstracts of Haryana.*

Figure 3.8
Yield of Main Non-Foodgrain Crops in Punjab and Haryana (Kg/Hectare) During the Current Period

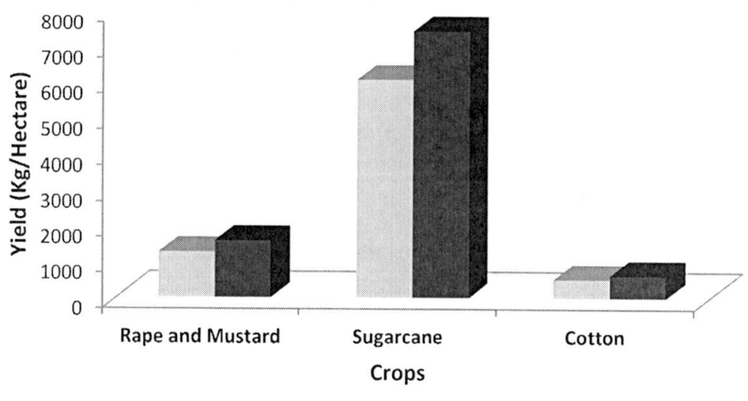

Next, the growth in yield of cotton crop is negative during the 1960s in both states. During the 1980s and 1990s, the yield of cotton increased in both states. During the 1990s the growth in yield of cotton remained almost the same in both states. Overall, yield of cotton increased more in Haryana than Punjab from 1966-67 to 2014-15.

Contribution of Punjab and Haryana in All India Foodgrains Procurement

In the preceding discussion, we came to know that since the period of the green revolution the production of rice and wheat continuously increased in Punjab and Haryana. These two states have made a significant contribution in the nation's total buffer stocks of foodgrains. In the total all India foodgrains procurement, Punjab's total share in rice and wheat procurement was 40 per cent and 46 per cent respectively.

Table 3.29

Percentage Share of Punjab and Haryana in All India Foodgrain Procurement (2015-16)

States	Rice	Wheat
Punjab	40	46
Haryana	13	29

Source: *Foodgrain Bulletin*, January 1, 2017 (Department of Food and Public Distribution, GoI).

On the other hand in case of Haryana, this state has a share of 13 per cent in rice procurement and 29 per cent in wheat procurement. Undoubtedly in these two states the procurement agencies as well as the network of agriculture markets are very effective compared to other states.

Livestock and Poultry in Punjab and Haryana

As already discussed Punjab since the colonial period has good breed milch cattles. During the green revolution periods, the government also tried to promote dairy farming as an additional source of income to farmers. At present in Punjab, there are 24.27 lakh cows and 51.59 lakh buffaloes. Out of these 0.60 lakh cows and 9.51 lakh buffaloes were milking.

Table 3.30

Livestock and Poultry in Punjab and Haryana (2012) (Lakhs)

Species	Punjab (i)	Percent-age	Haryana (ii)	Percent-age	Difference (i)-(ii)
Cows					
Total	24.27	100	18.08	100	6.19
In Milk	0.6	2.47	1.61	8.9	-1.01
Buffaloes					

Contd...

Total	51.59	100	60.85	100	-9.26
In Milk	9.51	18.43	20.02	32.9	-10.51
Birds (Poultry)	7.1	-	2.5	-	4.6

Source: *Statistical Abstracts of Punjab and Haryana* for Various Years.

On the other hand in Haryana, there were 18.08 lakh cows and 60.85 lakh buffaloes. Out of these, 1.61 lakh cows and 20.02 lakh buffaloes were in milk. It means in Haryana, a greater number of animals were in milk than Punjab. But it is quite interesting that the rural households, either landowners or landless, are now not interested to keep milch animals. In Haryana and Punjab this was found by many researchers[13]. The per capita availability of milk is the highest in Punjab, i.e. 1075 gms/day and in Haryana, it is 930 gms/day.

To conclude, both the states from the period of the pre-green revolution period to the current time, made remarkable progress in agriculture. The area under irrigation, and production of food grains increased in both states. But in Punjab agriculture developed more within this span than Haryana.

REFERENCES

1. For the changes in agrarian structure with the new agriculture technology in areas where the green revolution occurred we may see Francine Frankel, *India's Green Revolution: Economic Gains and Political Costs*, 1971.
2. With the increase in demand of agricultural labourers in Punjab and Haryana the migration of labourers from Bihar and Uttar Pradesh increased during the 1970s onwards. For details see Varinder Sharma, *Farm Workers of Punjab*, 2016.
3. For the growth in number of small farmers in Punjab (1970-2011), see Varinder Sharma, *Growth of Small Farmers in Punjab (1970-2011)*, Institute for Development and Communication (IDC), 2016.
4. For details, see D.P. Gupta and K.K. Shirigari, *Agricultural Development in Punjab*, 1980, p. 6.
5. Tubewell irrigation increased tremendously in Punjab and Haryana after 1966. For more details see Dhawan, *Development of Tubewell Irrigation in India*, 1982.

6. For the existing position of water tables in Indian states see Rodell Mathew et al., "Satellite Based Estimates of Ground Water Depletion in India", *Nature*, Vol. 460, 2009, pp. 999-1002.
7. For more details see S.S. Acharya and R.L. Jogi, *Farm Input Subsidies in Indian Agriculture*, Institute for Development Studies, 2004.
8. The debate on the employment of agriculture labourers (casual/ permanent labour) in agriculture started with the green revolution. Some scholars argued that the mechanization of farm operations is labour displacing. On the other side due to multiple cropping it is accepted that it increased the employment. For more details, see Hans P. Birwanger and Mark R. Rosenzweig: *Contractual Arrangements, Employment and Wages in Rural Labour Markets in Asia*, 1984 and H.S. Shergill: "Impact of New Technology on the Employment of Share Wage Animal Servants on Punjab Farms", *Indian Journal of Labour Economics*, Vol. 29, No. 4, 1987, pp. 72-81.
9. See S.S. Johl: *Problems and Prospects of Indian Agriculture* (Mimeo), pp. 6-7.
10. The debate on changing the existing cropping pattern which is mainly consisting of wheat and paddy in Punjab initiated by S.S. Johl and S.K. Raj: Future of Agriculture in Punjab, CRRID, 2002 and further this debate carried over by H.S. Shergill: *Diversification of Cropping Pattern: A Re-Examination*, IDC, 2006. The author concluded that wheat and paddy crops are beneficial to farmers and farmers will not immediately and easily shift from this established cropping pattern.
11. For details on suicides in rural Punjab, see Pramod Kumar, S.L. Sharma and Varinder: *Suicides in Rural Punjab*, Institute for Development and Communication (IDC), 2006.
12. For details see Guglani, Rakesh Kumar: *Hedging Mechanism for Farm Products Through Commodity Exchanges in India*, 2009, Department of Economics, Panjab University, Chandigarh.
13. For details, see Surinder S. Jodhka: "Emerging Ruralities: Revisting Village Life and Agrarian Change in Haryana", *Economic and Political Weekly*, Vol. 49, No. 26, 2014 and H.S. Shergill and Varinder Sharma: *Milk Production As Supplementary Source of Income for Rural Scheduled Caste Households in Punjab: Current Status and Strategy for Expansion and Viability*. Institute for Development and Communication, 2018.

CHAPTER 4

Agricultural Development and Equity

In the previous two chapters, we have discussed the process of agricultural development in two states, viz. Punjab and Haryana since the early period of the green revolution to the present time. This development which raised agriculture production and productivity in these two states was criticised by scholars on the ground of income inequality created by it between landowners and landless. The Marxist scholars criticised this development and argued that this development pauperised the small farmers and created a class of gentlemen farmers. Next, the landless labourers could not gain from this agricultural development and their wage rates almost stagnated over the years. There are no well defined indicators in development literature to measure inequality. In this chapter our aim is to capture the critical adverse impact of agricultural development on small farmers and landless labourers in Punjab and Haryana which created inequality. Section-I mainly focuses on the small farmers and Section-II on the landless agriculture labourers.

SECTION-I

Agricultural Development and Displacement of Small Farmers in Punjab and Haryana

A number of scholars using the Marxian framework hypothesised that the capitalist growth in agriculture in these two states deprived many small farmers and gradually the big farms emerged. Many scholars discussed this issue in the early phase of the green revolution[1]. Here, we have also discussed the growth of small farmers from 1971 to 2011 in Punjab and Haryana (Table 4.1).

Table 4.1

Growth of Small Farmers in Punjab and Haryana

States	Small & Marginal Farmers (Lakhs)			Area of Operational Holdings ('000 Acres)			Average Size of Holdings (Acres)		
	1970-71	2010-11	Growth Rate (1970-71 to 2010-11)	1970-71	2010-11	Growth Rate (1970-71 to 2010-11)	1970-71	2010-11	Change (ii)-(i)
Punjab	7.78 (56.54)	3.60 (34.18)	(-)1.91	1475 (15.03)	913 (9.32)	(-)1.19	1.90	2.54	0.64
Haryana	4.23 (46.27)	10.93 (67.59)	2.40	916 (10.74)	2036 (22.60)	2.02	2.15	1.85	(-)0.3

Note: Figures in brackets are percentages of all farmers.
Source: *Agriculture Census, (1970-71 and 2010-11)*, Ministry of Agriculture and Farmers' Welfare (Government of India).

In Table 4.1, the number of small farmers is given in Punjab and Haryana. Within forty years the number of small farmers in Punjab reduced by 1.91 lakhs. On the other hand in Haryana, the number of small farmers increased by 2.40 lakhs. The number of operational holdings of small farmers reduced by 1.19 per cent annually. In Haryana these holdings increased at the rate of 2.02 per cent annually. The average size of holdings of small farms increased by 0.64 acres in Punjab and in Haryana it reduced by 0.30 acres.

This type of growth process in these two states may be due to a number of reasons. The small farmers owing to non-profitability in agriculture are giving up agriculture[2] and their land is being purchased by big farmers. Secondly, due to rapid urbanisation, the rural area is shrinking[3]. Contrary to this in Haryana it seems the small farms have further sub-divided among farming households. With the result the number of small farmers increased.

Gain or Loss of Agriculture Land in Haryana

Table 4.2

Percentage of Households Out of Total Households Who Gained or Lost Land in Haryana: 1962-1972

Land Size Categories (Acres)	Gainers	Losers	No Change
0 – 2.5	13.73	64.59	21.68
2.5 – 5.0	16.01	57.89	26.10
5.0 – 10.0	21.64	25.79	52.57
10 – 15	28.10	31.44	39.96

Source: Sheila Bhalla: 'Changes in Acreage and Tenure Structure of Land Holdings in Haryana (1962-1972)', *Economic and Political Weekly*, Vol. 12, No. 1, Table 3, 1977, P.A.-4.

We have given the land lost and gained by farmers of different land categories either due to division, resumption by actual cultivators and eviction of tenants in Haryana. It is quite clear a proportion of households, i.e. around 29 per cent gained land. On the other hand, 49 per cent of households who have land above 5 acres are actual gainers of land. On the other hand, the maximum percentage of households who lost land belong to the land categories of 0-2.5 acres and 2.5-5.0 acres. About 91 per cent households in the land category of above 5 acres neither gained nor lost land. To conclude, immediately after the green revolution period owing to multiple reasons like resumption of land by actual land owners, eviction of tenants division of land, the small farmers mainly lost the land.

Access of Land to Scheduled Castes in Punjab and Haryana

Next, we have discussed the access of small farmers of Scheduled Castes to land in Punjab and Haryana (Table 4.3). Out of the total holdings of 10.53 lakhs in Punjab, the Scheduled Castes have 0.63 lakh holdings. It means only 5.9 per cent of holdings are owned by Scheduled Castes. In Haryana total holdings are 16.17 lakhs, and just 1.29 per cent of these holdings are with the Scheduled Castes. The numbers of Scheduled Caste small farmers are two times more in Punjab than Haryana. The average size of holdings of Scheduled Caste small farmers is 2 acres in Punjab and Haryana. It means the land is concentrated in the hands of a few castes in Punjab and Haryana.

Table 4.3

Access to Land of Scheduled Caste Farmers in Punjab and Haryana (2010-11)

States	Total Holdings (Lakhs)	Holdings of Scheduled Castes (Lakhs)	SC Small and Marginal Farmers (Lakhs)	Average Size of Holdings of SC Small and Marginal Farmers (Acres)
Punjab	10.53	0.63	0.40	2.00
Haryana	16.17	0.29	0.22	2.00

Source: *Agriculture Census, (1970-71 and 2010-11)* Ministry of Agriculture and Farmers Welfare (Government of India).

The caste composition of villages of Punjab and Haryana is different from the caste composition of other states of north India like Uttar Pradesh. In a village study it has been found in Punjab, that the village counting of 750 people has only 11 castes, whereas in Uttar Pradesh there are 22 castes. In Punjab and Haryana villages there is clear dominance of Jats. The state politics in both states is influenced by Jat peasantry. Further in Punjab and Haryana, a majority of farmers are Jats. Apart from Jats, other agriculturist castes are Rajputs, Sanis and Ahirs. But the land is monopolised by only these castes[4]. Similarly, most of the agricultural labourers are from Scheduled Castes. The main Scheduled Castes are Chamars (traditionally leather workers) and Sweepers/scavenger castes. The number of these Scheduled Castes varies from region to region in a state[5]. The land concentration seems to be highly skewed in Punjab and Haryana in the early period of the green revolution. In 1970-71, the Gini coefficient in Punjab and Haryana was around 0.50. Gradually, it came down and in 2010-11, it is 0.33 in Punjab and in Haryana it is 0.18. It means in Punjab the land is still concentrated in a few hands[6]. But it is interesting that there is no struggle among landless labourers for its distribution. During the 1960s in Punjab there was a struggle between tenants, landlords and landless labourers demanded land for agricultural purposes and building houses[7]. But with the passage of time, these struggles gradually lost steam. At present the main conflict is between land owners and the state at the time of land acquisition for various development projects or for setting up industries in the private sector[8].

In the next section, we have analysed the pattern of investment on small and large farmers in Punjab and Haryana.

Investment on Small and Big Farms in Punjab and Haryana

With the commercialisation of agriculture the fixed and variable capital investment on small and large farms increased continuously. The pattern of capital investment on small and large farms was even studied by some scholars in Punjab in the early period of the green revolution[9]. Here in Tables 4.4 and 4.5 we have given the investment per acre on small and large farms during the early period of the green revolution. The trend of investments on small farms in Punjab is not uniform from 1965 to 1972.

Table 4.4

Investment Per Acre on Small Farms in Punjab (Rs.)

Years	Tractors	Electric and Diesel Operated Tubewells	Chaff Cutters and Other Machinery	Total
1965	-	32.56	25.72	58.28
1966	-	3.92	3.21	7.13
1967	-	21.38	8.11	29.49
1968	-	19.23	8.24	27.47
1969	-	53.85	11.20	65.05
1970	-	98.77	13.85	112.62
1971	-	53.16	12.96	66.12
1972	12.20	66.01	27.79	106
Average Annual Investment from 1966	1.70	45.19	12.19	59.08

Source: B.D. Talib and A. Mazid: 'The Small Farmers of Punjab', *Economic and Political Weekly*, Vol. 11, No. 26, 1978, p. A-44, Table 5A.

The maximum investment was done during the 1970s and the minimum during 1966. The per acre investment on tractors remained the least, i.e. Rs. 1.70 and maximum was on electric and diesel operated tubewells. On chaff cutters and other machinery it remained around Rs. 12.19. During 1966, the eligibility criteria for tubewell loans for small farmers was relaxed by land mortgage banks and other agencies[10]. With the result, investment on electric and diesel engine operated tubewells on small farms increased in Punjab as already discussed in the previous chapter.

Table 4.5

Investment Per Acre on Big Farms in Punjab (Rs.)

Years	Tractors	Electric and Diesel operated Tubewells	Chaff Cutters and Other Machinery	Total
1965	74.47	60.64	19.73	154.84
1966	12.57	12.09	3.68	28.34
1967	-	23.21	13.53	36.74

Contd...

1968	86.56	36.27	26.67	149.5
1969	99.61	28.72	27.34	155.67
1970	131.26	25.73	22.49	179.48
1971	101.06	16.44	21.69	139.19
1972	58.03	48.55	36.33	142.91
Average Annual Investment from 1966	69.87	27.29	21.68	118.84

Source: B.D. Talib and A. Mazid: 'The Small Farmers of Punjab', *Economic and Political Weekly*, Vol. 11, No. 26, 1978, p. A-44, Table 5A.

Contrary to the investment on small farms, on the big farms the total investment is Rs. 118.84 per acre which is two times more than investment on small farms. On big farms, the maximum investment was on tractors, i.e. Rs. 69.87 per acre. On electric and diesel operated tubewells it remained Rs. 27.29 per acre and Rs.21.68 per acre on chaff cutters and other machinery. It seems the big farms became more mechanised than small farms in the early period of the green revolution.

Next, we have given the use of agriculture inputs by small and large farmers in recent times in Punjab and Haryana (Table 4.6).

Table 4.6

Use of Agriculture Inputs by Small and Large Farmers in Punjab and Haryana in the Current Time

Inputs	Punjab		Haryana	
	Small Farmers	Large Farmers	Small Farmers	Large Farmers
Percentage of Farmers with Tractors	48.55	85.80	46.77	74.50
Percentage of Farmers with Diesel Engine Pump Sets	19.20	38.33	21.17	25.61
Percentage of Farmers with Electric Pump Sets	44.07	76.20	34.72	52.26

Contd...

Percentage of Farmers Using Certified Seeds	74.41	83.26	90.50	96.96
Percentage of Farmers Using Soil Test Facilities	10.63	17.45	23.07	22.08

Source: Calculated from *All India Report on Input Survey (2011-12)*, Ministry of Agriculture and Farmers Welfare (Government of India).

At present in Punjab, 48.55 per cent of small farmers have tractors, whereas 85.80 per cent of large farmers own tractors. It means the big farms from the beginning of the green revolution to the present time are more mechanised than small farms[11].

Table 4.7

Characteristics of Small and Large Farms in Punjab and Haryana in the Current Time

Characteristics	**Punjab**		**Haryana**	
	Small Farmers	**Large Farmers**	**Small Farmers**	**Large Farmers**
Cropping Intensity	197	197	191	187
Percentage of Net Irrigated Area	98	99	90.2	91.6
Per Tractor Cultivated Net Sown Area (Acres)	5.00	15.00	4.00	17.00
Per Acre Use of Fertilisers for all Crops (Kg/Acre)	100	98	93	74
Percentage of Milking Animals Out of Total Milch Animals	48	50	40	42
Percentage of Holdings on Cash Tenancy	97	97	100	100

Source: Calculated from *All India Report on Input Survey (2011-12)*, Ministry of Agriculture and Farmers Welfare (Government of India).

On the other hand in Haryana, just 46.77 per cent of small farmers have tractors in Haryana and 74.50 per cent of large farmers own tractors. In Punjab 19.20 per cent of small farmers have diesel-operated tubewells and 44.07 per cent own power-operated tubewells. On the other hand, 38.33 per cent of large farmers use diesel-operated tubewells. In Haryana with small and large farmers the number of diesel-operated tubewells is the same. But the number of electric-operated tubewells is more with large farmers in Haryana. It means large farmers have more power-operated tubewells than small farmers in Punjab as well as in Haryana. Next, the percentage of farmers among large farmers who use certified seeds is more in Punjab compared to Haryana. Even the seeds replacement rate (SRR) is more among large farmers in Punjab[12]. The percentage of farmers using soil testing facilities is high of large farmers in Punjab as well as in Haryana. Undoubtedly this percentage is not satisfactory even among large farmers.

In the next sub-section we have described the main characteristics of small and large farms in Punjab and Haryana at present (Table 4.7).

Characteristics of Small and Large Farms in Punjab and Haryana

In the previous sections we have described the growth of small farmers in Punjab and Haryana. Under certain schemes and individual investments by farmers on the farms made the characteristics of small and large farms very peculiar. Here in Table 4.6 we have tried to show the characteristics of small and large farms in Punjab and Haryana. It is interesting to note that the cropping intensity of small and large farms is almost the same in Punjab and in Haryana. It is more on small farms and less on large farms. In Punjab, the percentage of net irrigated area is almost the same of small and large farms. Similarly, it is also the same in Haryana but less than Punjab.

In both states the per tractor cultivated net sown area is 5 acres in case of small farms and 15 acres in case of large farms. In Haryana, on small and large farms the difference in per tractor cultivated net sown area is 13 acres.

The per acre use of fertilisers in Punjab is almost the same on small and large farms. But in Haryana the use of fertilisers is more on small farms than large farms. It is quite interesting to note on small and large farms in Punjab as well as in Haryana, the percentage of milking animals is almost the same on small and large farms. Undoubtedly the herd size may differ on small and large farms in both states.

Finally, in both states on large and small farms, cash rent tenancy prevails to the same extent. After the green revolution, the reverse tenancy emerged in both states and the traditional share tenancy and share wage croppers got marginalised in both states. The small farmers were mainly deprived of the land lease market.

It means in Punjab and Haryana the characteristics of small and large farms over the years have become similar in many aspects.

The rapid commercialisation of agriculture in both states compelled the farmers to invest in agriculture. The farmers are always in need of short and long-term agriculture loans. The agricultural loans farmers avail of are from formal and informal financial institutions.

Prevalence of Agricultural Debt and Provision of Institutional Credit

From time to time the government published reports on the estimates of agriculture credit of the formal credit institutions. Some researchers also gave estimates of agriculture credit in Punjab and Haryana during the 1990s[13].

Table 4.8

Per Acre Debt in Punjab and Haryana (Rs.)

Type of Holdings	Punjab* (Rs.)	Haryana** (Rs.)
Small	10105	9065
Large	4230	5156
All Categories	5721	8246

Source: * H.S. Shergill (1998): *Rural Credit and Indebtedness in Punjab*, IDC, Chandigarh.

** T.R. Kundu, S. Kaushal and C.B. Sharma (1998): *A Study of Rural Indebtedness Amongst the Farmers in Haryana*, Department of Economics, Kurukshetra University, Haryana.

From Table 4.8, it is clear that in both states Punjab and Haryana the small farmers are under greater debt than large farmers. In both states on per acre basis small farms have two times more debt than large farms. Overall per acre debt in Haryana was Rs. 8246 and in Punjab it was Rs. 5721 in 1998. The farmers avail of credit from formal and informal credit agencies for daily domestic needs and agricultural purposes. At present the problem of mounting farm debt has become critical not only in these two states but also in all the other states of India. In a recent study the incidence of indebtedness among farm households has been estimated at 44 per cent in Punjab and 31 per cent in Haryana[14]. As per media reports the debt of 10 lakh farmers in Punjab is estimated at Rs. 90,000 crores and in Haryana, it pegged at Rs. 66,000 crores[15]. The farmers are continuously demanding of the waiving of loans in all the states. In Punjab, recently 38,000 marginal farmers received the relief of Rs. 209 crores[16].

The farmers' suicides are quite a debatable issue. The suicides of farmers, especially of small and marginal farmers, are linked to farm debt[17].

Here in Table 4.9 we have given the position of institutional credit used by small and large farmers.

Table 4.9

Institutional Credit Taken by Small and Large Farmers in Punjab and Haryana

Type of Credit	Punjab		Haryana	
	Small Farmers	Large Farmers	Small Farmers	Large Farmers
Total Number of Farmers Using Institutional Credit	216128 (60.08)	507789 (73.39)	479835 (44.23)	318540 (61.80)
Total Amount Taken (Crores)	3249	9286	4667	7194
Total Amount of Short-Term Loans (Crores)	766 (23.57)	5543 (59.69)	3268 (70.02)	5868 (81.56)
Share of Various Institutions in Short-Term Loans (Rs. Crores)				
Primary Agriculture Cooperative Societies (PACs)	730 (95.30)	5137 (92.67)	2385 (72.98)	4149 (70.71)

Contd...

Commercial Bank Branch (CBB)	5 (0.65)	80 (1.44)	129 (3.95)	181 (3.08)
Land Development Banks (LDB)	-	-	-	-
Regional Rural Bank Branch (RRBB)	31 (4.05)	326 (5.89)	754 (23.07)	1538 (26.21)
Total Amount of Long Term Loans (Crores)	2483 (76.42)	3743 (40.30)	1399 (29.97)	1326 (18.43)
Share of Various Institutions in Long Term Loan (Rs. Crores)				
Commercial Bank Branch (CBB)	1326 (53.40)	1863 (49.77)	693 (49.54)	602 (45.40)
Land Development Banks (LDB)	139 (5.60)	626 (16.73)	267 (19.08)	198 (14.93)
Regional Rural Bank Branch (RRBB)	1018 (41.00)	1254 (33.50)	439 (31.38)	526 (39.67)
Per Farmer Amount of Institutional Credit (Lakhs)	1.65	2.36	3.49	3.72
Per Acre Use of Institutional Credit (Lakhs)	0.39	0.69	0.61	0.92

Source: Calculated from *All India Report on Input Survey (2011-12)*, Ministry of Agriculture and Farmers' Welfare (Government of India).
Note: Figures in brackets are percentages.

If we look at Table 4.9 then there are just 60 per cent of small farmers in Punjab and 44 per cent in Haryana who avail of institutional credit from any institution. On the other hand in both states more than 60 per cent of farmers have access to institutional agricultural credit. In Punjab, large farmers avail of 88 per cent of total short-term loans and in Haryana 65 per cent of large farmers used short-term loans given by various formal financial institutions. The primary agricultural cooperative societies (PACS) provided maximum short-term loans for large and small farmers. In Punjab, 95 per cent of short-term loans are provided by PACS to small farmers and in case of large farmers 92 per cent of short-term loans are from PACS. The land development banks have no share in short-term loans, the contribution of regional rural banks in short-term loans like commercial banks is also not so high.

Similarly, in Haryana, the maximum short-term agriculture loans are provided by PACS. Here, the share of regional rural banks is much higher than Punjab in total short-term loans. If we compare the use of long-term loans by small and large farmers in these states, then in Punjab, out of the total amount of long-term loans, 40 per cent are availed of by small farmers. It is interesting that in Haryana, the small and large farmers availed of the long-term loans almost equally.

In Punjab and Haryana, commercial banks provided more than 40 per cent of the long-term loans to farmers. In Punjab, the land development banks provided just 5.60 per cent of long-term loans to small farmers and 19 per cent in Haryana. In both states the large farmers got more than 10 per cent of long-term loans.

In Punjab, the share of RRBs in total agriculture long-term loans is 40 per cent in case of small farmers and 33 per cent in case of large farmers. In Haryana, in case of small farmers it is 31 per cent and for large farmers it is 34 per cent.

The per farmer institutional credit was calculated around Rs. 1.65 lakhs in case of small farmers in Punjab and almost double in case of large farmers. But in Haryana, per farmer institutional credit was calculated around Rs. 3 lakhs.

In Punjab and Haryana, the per acre use of institutional credit is more on large farms and it is almost Rs.30,000 more than small farms in both states.

After discussing the availability of institutional credit among small and large farmers, we explained the annual income of small and large farmers in Punjab and Haryana. If we see Table 4.10 then it is clear the annual income of large farmers in both states is higher than small farmers.

The income from different sources highlights that income from cultivation is more for large farmers than small farmers. In Punjab income from cultivation of large farmers is three times more but in Haryana this ratio is seven. In both states the income from wages is more for small farmers. It is interesting that income from animal husbandry is more for large farmers in Punjab than for small farmers in Haryana. Again income from non-farming activities is more for small farmers in Punjab and Haryana. In case of Haryana, it was found that due to increasing unequal distribution of land, disparity in income distribution

Table 4.10

Weighted Mean Annual Income of Small and Large Farmers in Punjab and Haryana (Rs.)

Sources of Income	Punjab				Haryana			
	Small Farmers (i)	Large Farmers (ii)	Difference (ii)-(i)	Ratio (ii)/(i)	Small Farmers (i)	Large Farmers (ii)	Difference (ii)-(i)	Ratio (ii)/(i)
Cultivation	119322	399862	280540	3.35	48219	364087	315868	7.55
Wages	107481	46456	(-)61025	0.43	125818	16179	(-)109639	0.12
Animal Husbandry	26302	38428	12126	1.46	54510	36420	(-)18090	0.67
Non-Farming Activities	17006	10262	(-)6744	0.60	14888	522	(-)14366	0.04
Total Income	270113	495019	224906	1.83	243427	417210	173783	1.71

Source: Calculated from *Some Characteristics of Agricultural Households in India (NSS 70th Round), January-December, 2013.*

has increased from the early period of the 1960s to the 1980s. The present government is making efforts to double the farm income by 2022[18].

In the next section, we will explain how the agricultural labourers gained from the agricultural development in Punjab and Haryana.

SECTION-II

Agricultural Development and Status of Agricultural Labourers in Punjab and Haryana

The commercialisation of agriculture in Punjab and Haryana increased the demand of cash wage agriculture labourers for various farm operations. The migrant labourers also started coming to these states in search of work during the busy season in agriculture. Many migrant agriculture labourers have started living permanently in the villages of Punjab and Haryana and work as permanent farm servants[19].

The growth in the number of male agricultural labourers from 1971 to the current period is given in Appendix 29.

In Table 4.11, we have given the number of agricultural labourers in Punjab and Haryana and this is also depicted in Figure 4.1. In Punjab the total number of agricultural labourers is 15.88 lakhs and in Haryana number is 15.28 lakhs. It is interesting to note in both the states out of the total male workers, 15 per cent of workers work as agricultural labourers. Out of the total agricultural labourers, 78 per cent are males in Punjab and 68 per cent in Haryana. Further, it is important to note that out of the total agricultural labourers, 69 per cent are Scheduled Caste and 68 per cent are male from various Scheduled Castes in Punjab. But in Haryana, just 48 per cent are agricultural labourers from Scheduled Castes and the number of male Scheduled Caste agricultural labourers is 48 per cent. It means in Punjab a majority of Scheduled Caste households in rural areas are dependent on agriculture for employment. On the other hand, in Haryana, the Scheduled Caste households may be dependent on other occupations for their livelihood or they may have access to land for agriculture.

Table 4.11

Agricultural Labourers in Punjab and Haryana (2011)

States	**Total Agricultural Labourers (Lakhs) (i)**	**Male Agricultural Labourers (Lakhs) (ii)**	**Total Scheduled Caste Agricultural Labourers (Lakhs) (iii)**	**Male Scheduled Caste Agricultural Labourers (Lakhs) (iv)**
Punjab	15.88	12.39 (78) [15.35]	11.00 (69)	8.50 (68)
Haryana	15.28	10.41 (68) [15.30]	7.33 (48)	4.99 (48)

Note: (i) Figures in brackets are percentages.
(ii) In column (ii), in () brackets percentages are out of Total Agricultural Labourers.
(iii) In [] brackets percentages are out of Total Male Workers.
(iv) In column (iii) percentages are out of Total Agricultural Labourers.
(v) In column (iv) percentages are out of Total Male Agricultural Labourers.

Source: *Census of India.*

Figure 4.1

Agricultural Labourers in Punjab and Haryana

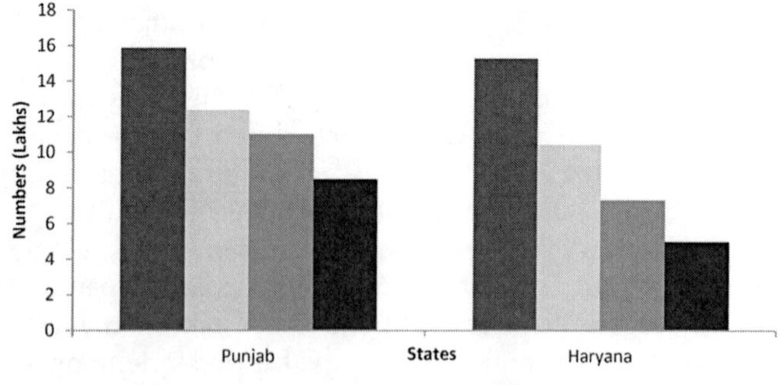

Table 4.12

Characteristics of Agriculture Labour Households in Punjab and Haryana

States	Estimated Number of Agriculture Labour Households ('000)		Family Size (Number of Members)		Percentage of All Agriculture Labour Households Owned Land	Percentage of SC Agriculture Labour Households Owned Land	Average Size of Land Holding (Acres)
	All Agriculture Labour Households	SC Agriculture Labour Households	All Agriculture Labour Households	SC Agriculture Labour Households			
Punjab	654	497 (76)	4.72	4.75	5.08	2.14	0.07
Haryana	351	261 (74)	4.86	4.65	12.25	2.18	0.84

Note: (i) Source: *General Characteristics of Rural Labour Households, Rural Labour Enquiry Report (66th Round of NSS 2009-10).*
(ii) Figures in brackets are percentages.

In the next sub-section we have given the number of agriculture labour households who have access to agriculture land and their family size etc.

In Table 4.12 and Figures 4.2A and 4.2B we have given the estimated number of agriculture labour households.

Figures 4.2A

Percentage of Agriculture Labour Households Owning Land in Punjab and Haryana

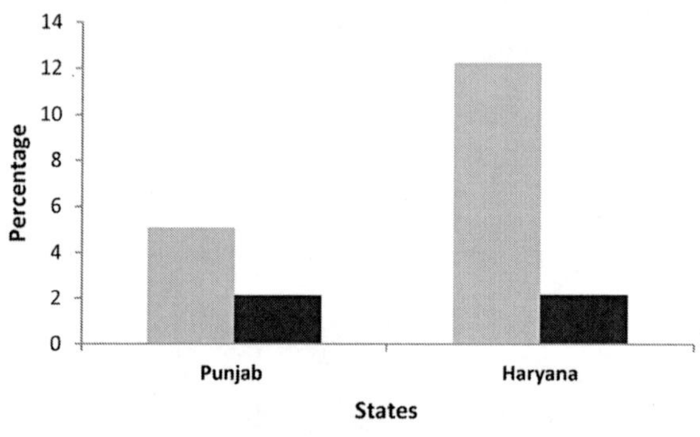

Figure 4.2B

Average Size of Land Holdings by Agriculture Labour Households in Punjab and Haryana

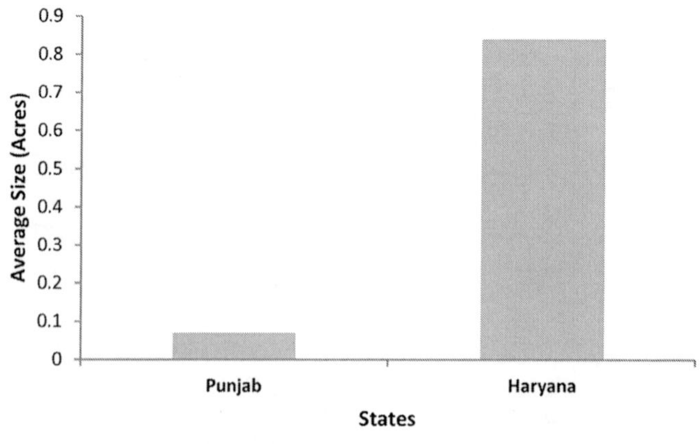

Out of the total estimated number of agriculture labour households in each state more than 70 per cent of agriculture labour households belong to Scheduled Castes. Overall in each state, i.e. Punjab and Haryana the number of family members are four on an average in agricultural labour households.

The distribution of land is highly skewed in both the states, between land owners and agricultural labour households. In Punjab only 5 per cent of agricultural labourers have land and only 2 per cent of Scheduled Caste agriculture labour households own land. The average size of their land holdings is less than one acre. On the other hand, these percentages are around 12 per cent but the size of their land holding is again less than one acre. It means in both states agricultural labour households have no access to agriculture land and the majority of them are landless. After discussing the ownership of land among agriculture labour households, we will elaborate how the wage rates of agricultural labourers increased with the agricultural development in Punjab and Haryana.

Agricultural Development and Wage Rates of Agricultural Labourers in Punjab and Haryana

The new agriculture technology completely changed the traditional labour markets in Punjab and Haryana[20]. With the commercialisation of agriculture a class of cash wage agricultural labourers emerged in these states. Many scholars, especially the leftist scholars, raised the issue of wage rates of agricultural labourers. The main concern of these scholars was: the new agriculture technology increased the production and productivity in agriculture. With the result income levels of land owners increased and the wage rates of landless agricultural labourers could not increase over the years. The money and real wage rates of agricultural labourers could not increase with the agricultural development[21].

Table 4.13

Wage Rate of Agricultural Labourers in Punjab and Haryana

States	Money Wage Rate (Rs./Day)		Money Wage Rate Index (Base: 1970-71)		Real Wage Rate (Rs./Day)		Real Wage Rate Index (Base: 1970-71)		CPI (Food) Agricultural Labourers	
	1970-71	2014-15	1970-71	2014-15	1970-71	2014-15	1970-71	2014-15	1970-71	2014-15
Punjab	6.39	317.66	100	4971	3.03	151	100	4983	100	2907
Haryana	6.64	371.50	100	5595	3.15	176.07	100	5589	100	2907

Note: (i) Money Wage Rates are of Male Agricultural Labourers at the ploughing season in the month of July.
(ii) Real Wage Rate was calculated on the base of CPI (Food) for Agricultural Labourers for the year 1970-71.
(iii) For the common base year we used the linking factor of CPI (Agricultural Labourers) given in *Indian Labour Journal*, February, 1996.

Sources: (i) *Indian Labour Journal* (various issues), Labour Bureau, Government of India.
(ii) *Statistical Abstracts of India*.

Figures 4.3A

Money Wage Rate of Agricultural Labourers in Punjab and Haryana

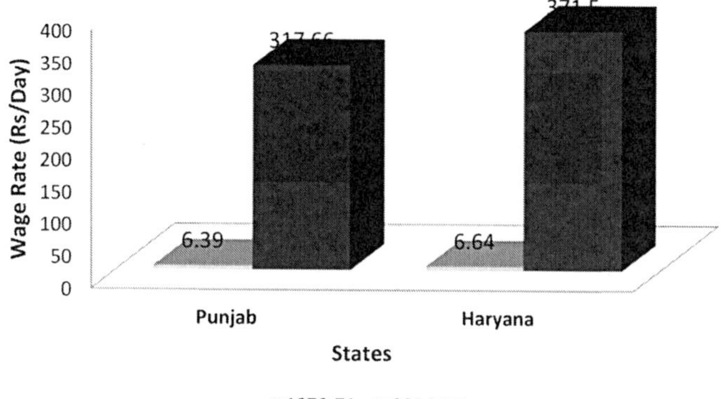

Figure 4.3B

Real Wage Rate of Agricultural Labourers in Punjab and Haryana

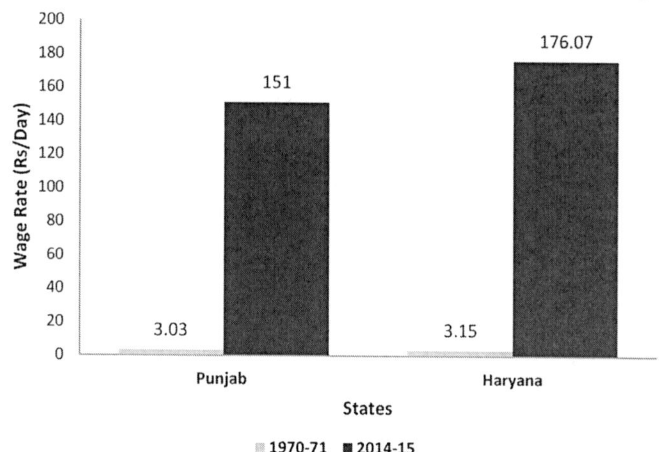

In Table 4.13 and Figures 4.3A and 4.3B we have given the money and real wage rates of agricultural labourers in Punjab and Haryana in the early green revolution period and current period. Further, the money wage rates, real wage rates and consumer price index of agricultural labourers from 1970-71 to 2014-15 are given in Appendices from 30 to 32. The trend analysis of the growth in money and real wage rates is given in Tables 4.14 to 4.16.

Table 4.14
Growth of Money Wage Rate of Agricultural Labourers in Punjab and Haryana

(Regression Analysis)
Equation: In (Y) = $\propto + \beta.T$
Y(Dependent Variable)= Money Wage Rate of Agricultural Labourers (Rs.)

States	Intercept ($\propto$)				Slope (β)				R^2			
	1970-71 to 1980-81	1981-82 to 1995-96	1996-97 to 2014-15	1970-71 to 2014-15	1970-71 to 1980- 81	1981-82 to 1995-96	1996-97 to 2014-15	1970-71 to 2014-15	1970-71 to 1980-81	1981-82 to 1995-96	1996-97 to 2014-15	1970-71 to 2014-15
Punjab	1.78	1.20	1.53	1.65	0.076 (20.98)***	0.11 (28.29)***	0.08 (13.38)***	0.09 (11.61)***	0.98	0.98	0.91	0.98
Haryana	1.78	1.57	1.51	1.58	0.07 (16.72)***	0.08 (8.30)***	0.09 (15.61)***	0.08 (45.69)***	0.97	0.84	0.93	0.98

Note: (i) Figures in brackets are t values.
(ii) t values are significant at: *** 1 per cent level.

Table 4.15

Annual Growth Rate in Money Wage Rate of Agricultural Labourers in Punjab and Haryana

States	1970-71 to 1980-81	1981-82 to 1995-96	1996-97 to 2014-15	1970-71 to 2014-15
Punjab	7.00	11.00	8.00	9.00
Haryana	7.00	8.00	9.00	8.00

Now, first look at Table 4.13, during 1970-71, the money wage rate of agriculture labourers was almost the same in both states, but during 2014-15, it increased in both states and it is marginally higher in Haryana. The money wage rate index shows money wage rates increased in both states and it remained higher in Haryana than Punjab. The real wage rates are half of the money wage rates in both states during 2014-15 compared to 1970-71. The real wage rate index is almost the same as the money wage rate index. The consumer price index is almost the same in Punjab and Haryana.

Further the growth in money and real wage rates in Punjab and Haryana is given in Tables 4.14 and 4.16. From Tables 4.14 and 4.15, it is clear that during the period 1970-71 to 1980-81 the money wage rates increased by 7 per cent per annum. In the next time period, i.e. from 1981-82 to 1995-96, in Punjab the money wage rates increased by 11 per cent per annum but in Haryana it remained around 8 per cent. The money wage rates during the period 1996-97 to 2014-15 and the growth rates reduced in Punjab and marginally increased in Haryana. Lastly, from 1970-71 to 2014-15, the money wage rates in Punjab increased by 9 per cent per annum, and in Haryana it increased by 8 per cent. From the regression analysis it is clear that the money wage rates of casual agricultural labourers in Punjab and Haryana did not increase continuously. During all the time periods, especially from 1970-71 to 1980-81 and 1996-97 to 2014-15, the growth rates remained almost the same in Punjab and Haryana. This stagnant growth may be due to the fast mechanisation of agricultural operations resulting in reduced demand of labourers in agriculture. Secondly, it may be due to oversupply of agricultural labourers in the village labour market, owing to which agricultural labourers can not bargain with land owners for higher wages.

Table 4.16

Growth of Real Wage Rate of Agricultural Labourers in Punjab and Haryana

(Regression Analysis)

Equation: In (Y) = ∝ + β.T

Y(Dependent Variable)= Money Wage Rate of Agricultural Labourers (Rs.)

States	Intercept (∝)				Slope (β)				R^2			
	1970-71 to 1980-81	1981-82 to 1995-96	1996-97 to 2014-15	1970-71 to 2014-15	1970-71 to 1980- 81	1981-82 to 1995-96	1996-97 to 2014-15	1970-71 to 2014-15	1970-71 to 1980-81	1981-82 to 1995-96	1996-97 to 2014-15	1970-71 to 2014-15
Punjab	1.03	0.45	0.78	0.90	0.07 (21.18)***	0.11 (28.29)***	0.09 (13.37)***	0.09 (58.79)***	0.98	0.98	0.91	0.98
Haryana	1.03	0.82	2.82	0.83	0.07 (16.68)***	0.08 (8.30)***	0.09 (16.43)***	0.08 (45.69)***	0.97	0.84	0.93	0.98
Number of Years	10	15	19	44	-	-	-	-	-	-	-	-

Note: (i) Figures in brackets are t values.

(ii) t values are significant at: *** 1 per cent level.

Table 4.17

Annual Growth Rate in Real Wage Rate of Agricultural Labourers in Punjab and Haryana

States	1970-71 to 1980-81	1981-82 to 1995-96	1996-97 to 2014-15	1970-71 to 2014-15
Punjab	7.00	11.00	9.00	9.00
Haryana	7.00	8.00	9.00	8.00

Next, if we look at Tables 4.16 and 4.17 then the growth in the real wage rates of agricultural labourers is almost equal to the money wage rates. It means real and money wage rates increased at the same pace in both states. But in quantum the real wage rates almost remained half than the money wage rate from the early green revolution period to the current period (see Table 4.13).

To sum up, most of the agricultural labourers are landless and belong to Scheduled Castes and no massive growth happened in their real and money wage rates since the period of the green revolution. Next in Table 4.18 and Figures 4.4(A) and 4.4(B) we have given the annual per capita total consumption expenditure of agriculture labour households in Punjab and Haryana.

Table 4.18

Annual Per Capita Total Consumption Expenditure of Agricultural Labour Households in Punjab and Haryana

States	Annual Per Capita Expenditure (Rs.)	Percentage Consumption Expenditure on Food Items Out of Total Expenditure	Percentage Consumption Expenditure on Non-Food Items Out of Total Expenditure
Punjab	12015.48	54.5	45.5
Haryana	9684.24	60.1	39.9
National Average	8748.40	57.2	42.8

Source: Consumption Expenditure of Rural Labour Households (66th Round of NSS), 2009-10, *Rural Labour Enquiry Report*.

Figures 4.4A

Annual Per Capita Expenditure of Agricultural Labour Households in Punjab and Haryana

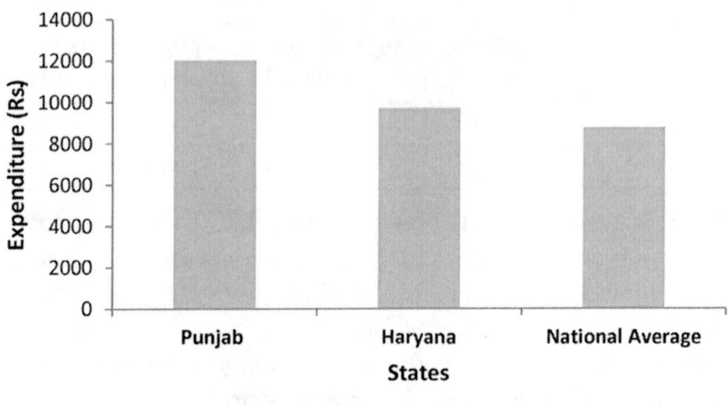

Figure 4.4B

Percentage Consumption Expenditure on Food and Non-Food Items in Punjab and Haryana

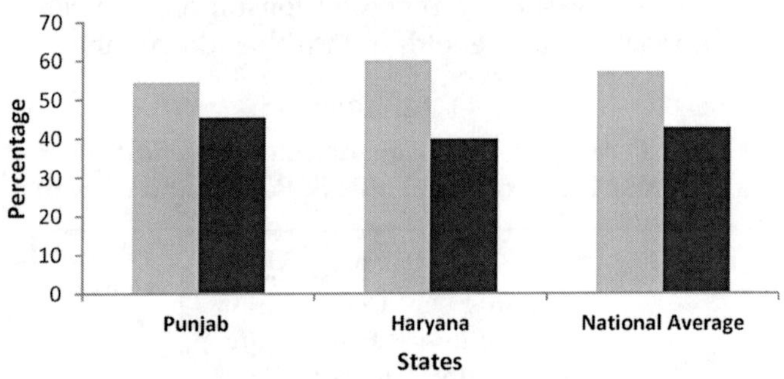

Annual Per Capita Consumption Expenditure in Punjab and Haryana

In this section and in Table 4.18 and Figures 4.4A and 4.4B, we will explain their composition of consumption expenditure in Punjab and Haryana. The annual per capita expenditure in Punjab is around Rs. 12,000 which is higher by Rs. 4000 compared to national average. In Haryana this annual per capita consumption is Rs. 9684.24 which is higher by Rs. 1000 relatively to national average but less by Rs. 3000 compared to Punjab. Out of the total consumption expenditure half of the expenditure is on food items in Punjab but in Haryana around 60 per cent of the expenditure is on food items.

If we compare the expenditure on non-food items then around 40 per cent of the total consumption expenditure is on non-food items in both Punjab and Haryana. It means a major portion of total consumption expenditure of agriculture labour households goes only on food items.

In the next sub-section we will explain the level of debt of agriculture labour households in Punjab and Haryana. In literature, it is accepted that in agriculture the labour contracts are interlinked due to the debt of agriculture labour households[22].

Table 4.19

Extent of Debt of Agricultural Labour Households in Punjab and Haryana

States	Percentage of Indebted Households	Extent of Indebtedness (Rs.)		Nature of Loan	
		Average Debt Per Household	Average Debt Per Indebted Household	Hereditary (%)	Contracted (%)
Punjab	42.90	15312	35722	2.44	97.56
Haryana	42.90	22560	52603	3.81	96.19

Source: *Rural Labour Enquiry Report on Indebtedness Among Rural Labour Households (66th Round of NSS), 2009-10.*

If we look at Table 4.19 then we can see that the percentage of indebted agriculture labour households is around 40 per cent in Punjab and Haryana. The average per household debt in Haryana is one and a half times more than Punjab. But the average debt per indebted household is Rs. 35,722 in Punjab and Rs. 52,603 in Haryana. The average debt per indebted household is almost one and a half times more than the average debt per household in both states. It is interesting that more than 90 per cent of debt in both states is due to labour contracts with the employers.

Next in Table 4.20, the extent of debt of Scheduled Caste agricultural labour households in Punjab and Haryana has been given. In both states around forty per cent of the total agriculture labour Scheduled Caste households are in debt in Punjab and Haryana. The average debt per Scheduled Caste agriculture labour household is Rs. 10,785 and Rs. 16,787 in Haryana. The average debt per indebted household is almost double of the average debt per Scheduled Caste agriculture labour households. In both states 90 per cent of the debt of the Scheduled Caste agriculture labour households arises due to their labour contracts. It means a majority of the agricultural labourers take an advance from their employers (farmers) and at the end of the agriculture season they return the loans. Those who do not return the loans fall into the 'debt trap' and become bonded labourers.

Table 4.20

Extent of Debt of Scheduled Caste Agricultural Labour Households in Punjab and Haryana

States	Percentage of Indebted Households	Extent of Indebtedness (Rs.)		Nature of Loan	
		Average Debt Per Household	Average Debt Per Indebted Household	Hereditary (%)	Contracted (%)
Punjab	44.80	10785	24078	1.54	98.46
Haryana	41.40	16787	40539	5.72	94.28

Source: *Rural Labour Enquiry Report on Indebtedness Among Rural Labour Households (66th Round of NSS), 2009-10.*

Figure 4.5A

Percentage of Indebted Agricultural Labour Households in Punjab and Haryana

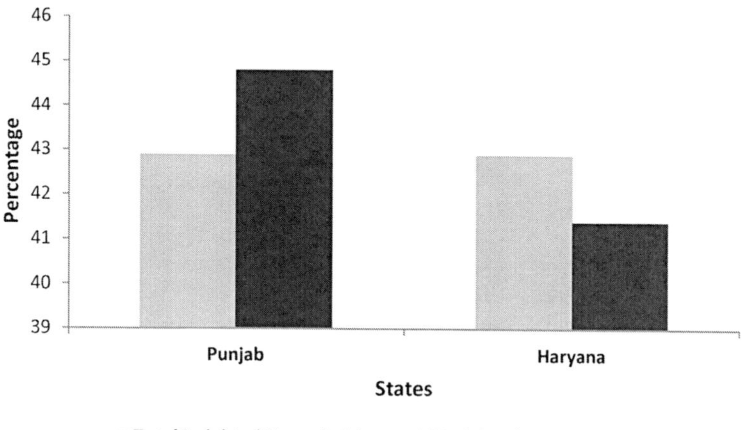

Figure 4.5B

Percentage of Indebted Agricultural Labour Households in Punjab and Haryana

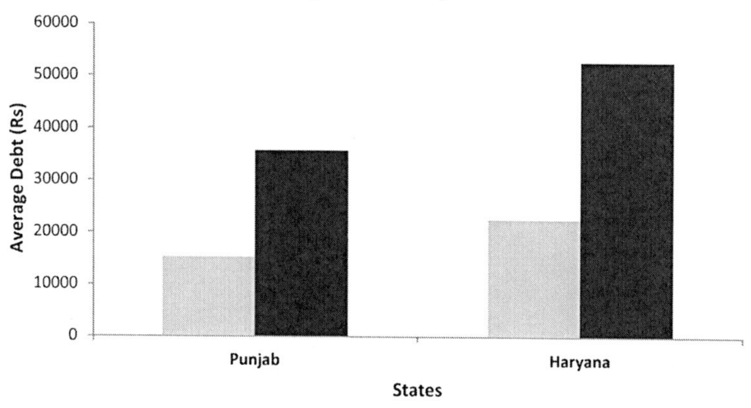

Figure 4.5C

Percentage of Indebted Agricultural Labour Households in Punjab and Haryana

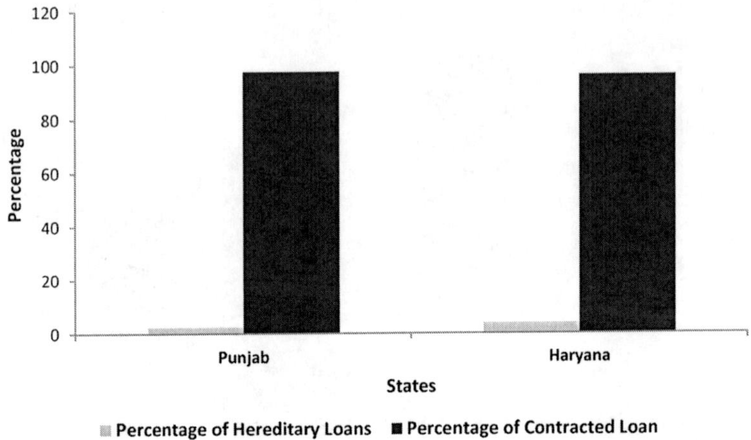

Figure 4.6A

Extent of Debt of Scheduled Caste Agricultural Labour Households in Punjab and Haryana

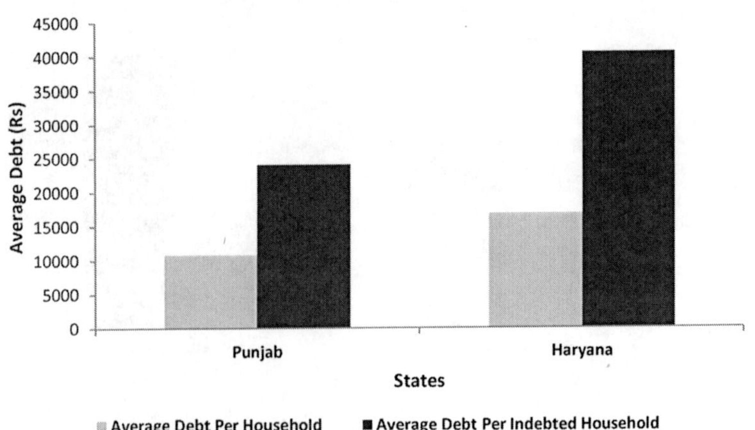

Figure 4.6B

Nature of Loan of Scheduled Caste Agricultural Labour Households in Punjab and Haryana

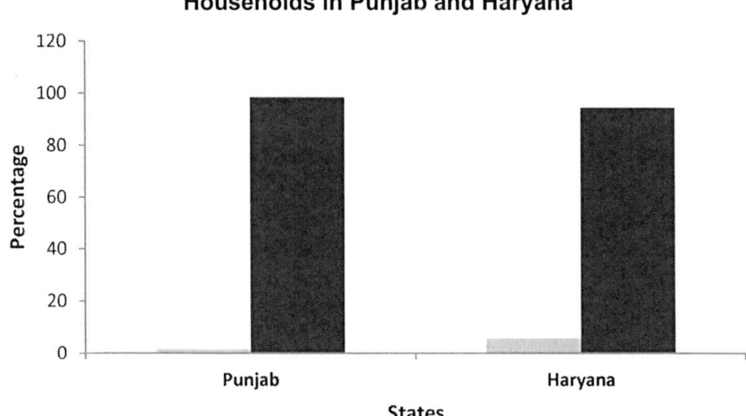

Percentage of Hereditary Loans ■ Percentage of Contracted Loan

To sum up from the preceding discussion, the rapid agricultural development in these two states has not benefited the small and marginal farmers and the wage rate of agricultural labourers to a large extent, especially real wage rates has not increased at a fast pace. The income levels only of the large farmers increased with this agricultural development.

REFERENCES

1. The agricultural development in Punjab and Haryana was linked with the development of capitalism in agriculture. Many scholars also concluded that the development of capitalism in agriculture pauperised the small peasants and big farmers or absentee landlords purchased their lands. For details, see A. Majidi and B.D. Talib: "The Small Farmers of Punjab", *Economic and Political Weekly*, Vol. 11, No. 26, pp. A42-A46, 1976.
2. For details, see Sukhpal Singh and Shruti Bhogal: "Depeasantisation in Punjab: Status of Farming Who Left Farming", *Current Science*, Vol. 106, No. 10, pp. 1364-1381, 2014.
3. See Varinder Sharma: *Growth of Small Farmers in Punjab: 1971 to 2011*, Institute for Development and Communication, 2017.
4. For details see Joseph E. Schwartzberg, "Caste Regions of the North India Plain", in *Structure and Change in Indian Society* (eds.) Milton Singer and Bernard Cohn, 1968, pp. 99-100.

5. Ibid., p. 100.
6. See Appendix 33, S.Nos. 34 and 36.
7. For more details on agricultural workers' struggle in Punjab for the agriculture land and land sites for housing see Master Hari Singh: *Agricultural Workers' Struggle in Punjab*, 1980.
8. For more details, see Sucha Singh Gill: "Land Acquisition in Liberalised Indian Economy: Changing Nature and Contentious Issues", in *Dynamics of Rural Transformation in India* by M.R. Khurana (ed.), 2018.
9. See A. Majidi and B.D. Talib: "The Small Farmers of Punjab", *Economic and Political Weekly*, Vol. 11, No. 26, pp. A42-A46.
10. The rapid expansion of institutional credit and electrification and drop in criteria of minimum holdings for the tubewell loans gradually increased the number of tubewells in Punjab and Haryana which slowly disappeared. See B.D. Dhawan: *Development of Tubewell Irrigation in India*, 1982.
11. The large farms are always more mechanised than small farms. See Ashok Rudra: "Employment Patterns in Large Farms of Punjab", *Economic and Political Weekly*, Vol. 6, No. 26, 1971, pp. A89-A94.
12. The farmers lack knowledge of new seed varieties and have less access to new varieties too. In one study, it was found that the seed replacement rate in case of paddy crop is just 18.01 per cent in case of small farmers and 31.48 per cent among large farmers. For details, see M.S. Sidhu et al: "Sources, Replacement and Management of Paddy Seeds by Farmers in Punjab", *Agricultural Economics Research Review*, Vol. 22, Issue 2, 2009, pp. 323-328.
13. The sample-based studies estimated per acre institutional and non-institutional agricultural debt in Punjab and Haryana. For details, see H.S. Shergill: *Rural Credit and Indebtedness in Punjab*, IDC, 1998 and T.R. Kundu, S. Kaushik and C.B. Sharma: *A Study of Rural Indebtedness Among the Farmers in Haryana*, Department of Economics, Kurukshetra University, Haryana, 1998.
14. *All India Rural Financial Inclusion Survey (2016-17)*, NABARD, 2018.
15. *The Hindustan Times*, June 12, 2017.
16. *The Economic Times*, June 13, 2018.
17. For details see Pramod Kumar, S.L. Sharma and Varinder Sharma: *Suicides in Rural Punjab*, Institute for Development and Communication (IDC), 2006 and Kesar Singh Bhangoo,

Lakhwinder Singh and Rakesh Sharma: *Agrarian Distress and Farmer Suicides in North India*, 2016.

18. For details, see Satya Paul: "Green Revolution and Income Distribution Among Farm Families in Haryana, 1969-70 to 1982-83", *Economic and Political Weekly*, Vol. 24, No. 51-52, pp. A154-A158. See Ramesh Chand: "Why Doubling Farmers' Income by 2022 is Possible", *Indian Express*, April 4, 2016.
19. S.S. Johl: *Gains of the Green Revolution: How They Have Been Shared in Punjab*, Department of Economics & Sociology, 1975, Punjab Agricultural University, Ludhiana.
20. Traditionally, the agricultural labourers in united Punjab were paid in kind at the time of harvesting but with commercialisation of agriculture the kind wages was replaced with cash wages. For details, see Varinder Sharma: *Farm Workers of Punjab*, 2016.
21. A number of scholars analysed the wage rates of casual agricultural operations after the green revolution period. It has been concluded that the money and real wage rates of casual agriculture labourers marginally increased but no continuous rise came. For details, see Sheila Bhalla: "Real Wage Rates of Agricultural Labourers in Punjab: 1961-1977, A Preliminary Analysis", *Economic and Political Weekly*, Vol. 14, No. 26, pp. A57-A68, 1979 and S. Mahendra Dev: Poverty and Employment: *Roles of Agriculture and Non-Agriculture*, V.V. Giri Memorial Lecture, 58th Annual Conference, Indian Society of Labour Economics, IIT, Guwahati, 2016.
22. For such labour contracts we may see Pranab Bardhan and Ashok Rudra: Terms and Conditions of Labour Contracts in Agriculture, Results of a Survey in West Bengal 1979, *Bulletin of the Oxford University of Institute of Economics & Statistics*, Vol. 43, No. 1, pp. 89-111, 1981 and Clive Bell and T.N. Srinivasan: "Interlinked Transactions in Rural Markets: An Empirical Study of Andhra Pradesh, Bihar and Punjab", *Oxford Bulletin of Economics and Statistics*, Vol. 51, No. 1, pp. 73-83, 1989.

CHAPTER 5

Summary and Conclusions

To collate together the arguments and findings of Chapter-I through Chapter-IV, we illustrate in the first section below the emergence of two states, Punjab and Haryana, as a hub of new agriculture technology. In the next sections, we discuss the scenario of agricultural development from the early period of the green revolution to the present time. Finally, we explain how this agricultural development supported the small farmers and landless agriculture labourers.

I. Regions of New Agriculture Technology: Punjab and Haryana

In August 1947, British India was partitioned into Pakistan and India. With this tragic partition, West Punjab became part of Pakistan. East Punjab integrated with India. The partitioned portion of Indian Punjab received 47 per cent of the population, 34 per cent of the area and 20 per cent of the canal irrigated area (Chapter-I). The economy of East Punjab was shattered at the time of partition. In East Punjab which comprised Punjab and Haryana, industry was not developed. Besides, most of, whatever little industry that had developed by then had been damaged during communal riots at the time of partition. The agriculture was also not so developed and most of the area faced food deficit.

After partition, again in 1966 the state of Punjab was reorganised on the basis of language into the present states of Punjab and Haryana and the hilly region of the state was incorporated in Himachal Pradesh. Immediately after the re-organisation of these states, the process of formulation and execution of plans to develop the economies started. The prime task before the state government was to wipe out the food deficit. Large tracts of the state did not have irrigation network. The scenario in foodgrain production was becoming grimmer in the country. Further, the thesis of the Paddock brothers on the deteriorating food security in India and in other food deficit nations was an alarm for planners and politicians. At this juncture when the population was increasing the deficit in foodgrain production was becoming increasingly larger. Therefore, in the early three Five Year Plans the main focus remained on enhancing the agricultural output to support rapid industrialisation and the social commitment was to reduce economic disparities in rural areas. Later on, in 1965, the programme for modernisation of agriculture in the irrigated area, with introduction of HYV (high-yielding varieties) of seeds was initiated. This modernisation process became popularly known as the "Green Revolution". The states of Punjab and Haryana adopted this modernisation process during the 1960s in a big way. The agriculture production and productivity, especially of wheat and rice increased tremendously in these two states. Not only in India, but all over the world, Punjab and Haryana became successful examples of the Green Revolution.

II. Agricultural Development on the Eve of the Green Revolution in Punjab and Haryana

On the eve of the green revolution the contribution of agriculture and animal husbandry in total net state domestic product of Punjab and Haryana was around 50 per cent. On agriculture directly or indirectly 19.36 lakhs of agriculture workers (cultivators and agriculture labourers) were dependent in Punjab and 20.37 lakhs in Haryana. In both states there was around 15 per cent of the uncultivated area during the 1960s. The percentage of net irrigated area to net sown area was around 58 per cent in Punjab and around 35 per cent in Haryana. In both states the main source of irrigation was canals and there was potential to develop the tubewells. It seems (Chapter-2 and Table 2.6) the agriculture infrastructure was more advanced in Punjab than Haryana. The use of fertilisers was three times more in Punjab than Haryana. Similarly, the number of tubewells was three times more in Punjab than Haryana. The consumption of electricity in agriculture was two times more in Punjab than Haryana. The agricultural markets in Haryana during the 1960s were almost non-existent but in Punjab this number was around eighty-four. In Punjab loans advanced by these primary agriculture cooperative credit societies was around Rs. 25 crores which was Rs. 15 crores more than Haryana.

In both the states, in the early period of the green revolution, the main foodgrain crops were wheat, bajra and gram (Chapter-2, Table 2.7). In Punjab around 50 per cent of area was under wheat cultivation. In Haryana, along with wheat, gram and bajra were other major foodgrain crops.

Among the non-foodgrain crops, cotton was the leading crop in both states. This crop was cultivated on more than 40 per cent of total area under non-foodgrain crops in Punjab and 38 per cent in Haryana. In Haryana, rape and mustard was the main non-foodgrain crop, it was cultivated on thirty per cent of the total area under non-foodgrain crops.

Since the colonial period the farmers in Punjab owned high quality breed milch animals. In both states the number of buffaloes was three times more than cows (Chapter-2, Table 2.13). The landless and landowners are now not interested to keep milch animals in both states due to non-profitability and other reasons. The agriculture was quite developed in Punjab in terms of net sown area, irrigation and agriculture machinery. The institutional infrastructure like cooperative credit was also advanced in Punjab. The area, production and yield of foodgrain and non-foodgrain crops was also higher in Punjab than Haryana.

III. Agricultural Development in the Post Green Revolution Period in Punjab and Haryana

The contribution of agriculture and animal husbandry in total net state domestic product continuously declined in Punjab and Haryana since the 1960s. At present this share in Punjab is 29.28 per cent and in Haryana it is 19.95 per cent (Chapter-3, Table 3.4). On agriculture, 35.28 lakh agriculture workers (cultivators and agricultural labourers) depend for their livelihood in Punjab and in Haryana this figure is 40.08 lakhs. From 1961-2011, this number in Punjab increased by 15.86 lakhs and in Haryana by 19.71 lakhs.

The percentage of net irrigated area out of the net sown area touches 99 per cent in Punjab and in Haryana, it is around 85 per cent. In the early green revolution period the maximum area was irrigated by canals, but gradually the tubewells became the major source of irrigation in both the states. The tubewell irrigation developed fast in the two states owing to land consolidation, easy availability of institutional credit and rural electrification. Moreover, the individual investment by farmers in establishing tubewells sets in both states is remarkable. At present, it is also an issue of debate with tubewell irrigation, the use of groundwater increased tremendously and with this the water table is falling in both states.

The use of other main agriculture inputs like fertilisers, tractors and electric tubewells increased tremendously in both states (Chapter-3, Table 3.8).

The area under rice has increased by almost 5 per cent per annum from 1966 onwards. In the early period of the green revolution this crop was not cultivated on a large scale in both states. The area under rice increased in both states due to development in irrigation facilities and assured market for this crop at the minimum support price. At present in the central pool of foodgrain procurement, the share of Punjab is 40 per cent in rice and 46 per cent in wheat. On the other hand, the share of Haryana is 13 and 29 per cent respectively in these two states.

To conclude, in the post green revolution period the infrastructure in agriculture in both states increased very fast. The number of tractors, tubewells and facilities of institutional credit increased tremendously since the 1960s. Rice and wheat became the major crops in both states after the green revolution.

IV. Agricultural Development and Equity

In the previous sections, we discussed how agricultural development took place in these two states. At present these two states are the pillars of India's food security. Now the question arises how this agricultural development benefited the small farmers and landless agricultural labourers. During the 1960s among development theorists and planners, it was the main concern to raise overall output in the economy. They believed that such an increase in output would "trickle down" and benefit lower sections of society. Contrary to this, during the 1970s, the rise in per capita gross national product has not produced a downward trend in poverty, income inequality and increased employment.

A particular section got benefits of this development. Still, there is no consensus on the definition or criteria of 'equity'. Rather, criteria is defined for the equity in society like consumption by the poor, infant mortality, literacy rates, dispersion in incomes and ownership of productive resources. Here, within the preview of this theoretical framework, we analysed how agricultural development in Punjab and Haryana benefited small farmers and landless agricultural labourers. In literature, scholars accepted that the new agriculture technology increased the income levels of all farmers. But the main gains remained in favour of the large farmers. Many scholars within the Marxian framework also concluded the agricultural development pauperised the small farmers. The small farmers' land was purchased by large farmers. In Punjab from 1971 to 2011 onwards the number of small farmers decreased by 1.91 per cent per annum. The average size of land holding has increased by 0.64 acres. It means the number of big farms is increasing in Punjab. The small farmers due to multiple reasons like non-profitability, urbanisation and shifting towards other occupations are moving away from agriculture (Chapter-4, Table 4.1). Contrary to this in Haryana the number of small farmers is still increasing which shows still more dependency on agriculture.

The Scheduled Castes are still deprived of agricultural land in Punjab and Haryana. In Punjab only 5 per cent of the total holdings belong to Scheduled Castes and in Haryana this figure stands at 2 per cent. In both states the dominant caste of cultivators is the Jats. The other land owning castes are Sainis, Rajputs and Ahirs. The cash tenancy in both states replaced the share wage croppers and share wage permanent farm servants.

At the beginning of the green revolution the per acre investment was higher on large farms than small farms (Chapter-IV and Tables 4.4 and 4.5). Still, the agricultural inputs use and ownership of agriculture machinery is higher on the large farms (Table 4.8, Chapter-IV). The debt on the small farms is also higher in both states. The per acre debt on small farms is estimated at Rs.10105 in Punjab and in Haryana it remained Rs.9065 (Chapter-IV and Table 4.8). The share in the institutional credit of small and large farmers has a huge gap. In Punjab just 60 per cent of small farmers use institutional credit and in Haryana this figure is 44.23 per cent (Chapter-4 and Table 4.9).

The gap in annual income from cultivation of small and large farmers is of Rs. 2.80 lakhs in Punjab and Rs. 3.16 lakhs in Haryana (Chapter-IV and Table 4.10).

Finally, we have discussed the status of agriculture labourers with this agricultural development. In Punjab the number of landless agricultural labourers is 15.88 lakhs and in Haryana this figure touches 15.28 lakhs. Out of these, 78 per cent are males in Punjab and in Haryana this figure is 68 per cent. In Punjab, 69 agricultural labourers are Scheduled Caste and in Haryana this figure is 48 per cent. A majority of Scheduled Caste agriculture labourers are male, i.e. 68 per cent in Punjab and 48 per cent in Haryana (Chapter-IV and Table 4.11).

Next, a debate on the money and real wage rate of landless agriculture labourers is continuously going on since the period of the green revolution. The money wage rates of casual agricultural labourers in Punjab and Haryana did not continuously increase. From 1970-71 to 1980-81, the wage rate almost increased by 7 per cent annually. During 1981-82 to 1995-96 in Punjab the growth rate remained 11 per cent per annum and in Haryana it was just 8 per cent. In the next period, i.e. from 1996-97 to 2014-15, the wage rate declined by 3 per cent compared to the previous period and in Haryana it increased marginally by 1 per cent. Overall the money wage rate has increased by 9 per cent annually in Punjab and 8 per cent in Haryana from 1970-71 to 2014-15.

The real wage rate in Punjab and Haryana increased almost at the same rate as the money wage rate increased. But the quantum of real wage rate remained almost half of the money wage rate (Table 4.1 and Table 4.17, Chapter-4).

A major portion of their wages, i.e. 54 per cent in Punjab and 60 per cent in Haryana is spent on food items. Almost 40 per cent of agriculture labour households are under debt in Punjab and Haryana (Tables 4.18 and 4.20, Chapter-IV).

In a nutshell, the main gains of this agricultural development remained and went to large farmers. Landless agricultural labourers and Scheduled Castes have not benefited much.

APPENDIX A

Maps

MAP 1.1

AREA OF THE STUDY

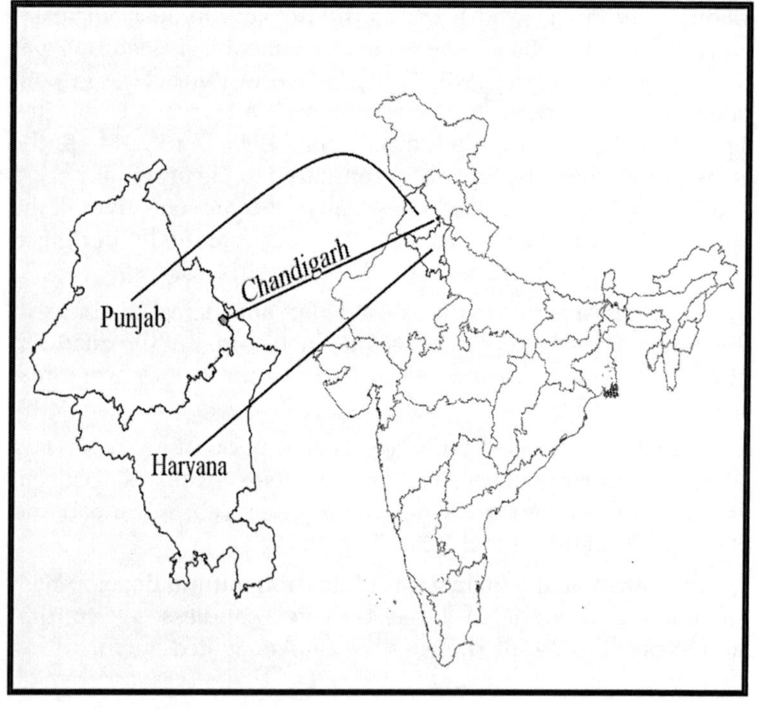

MAP 1.2

PUNJAB ADMINISTRATIVE DIVISION

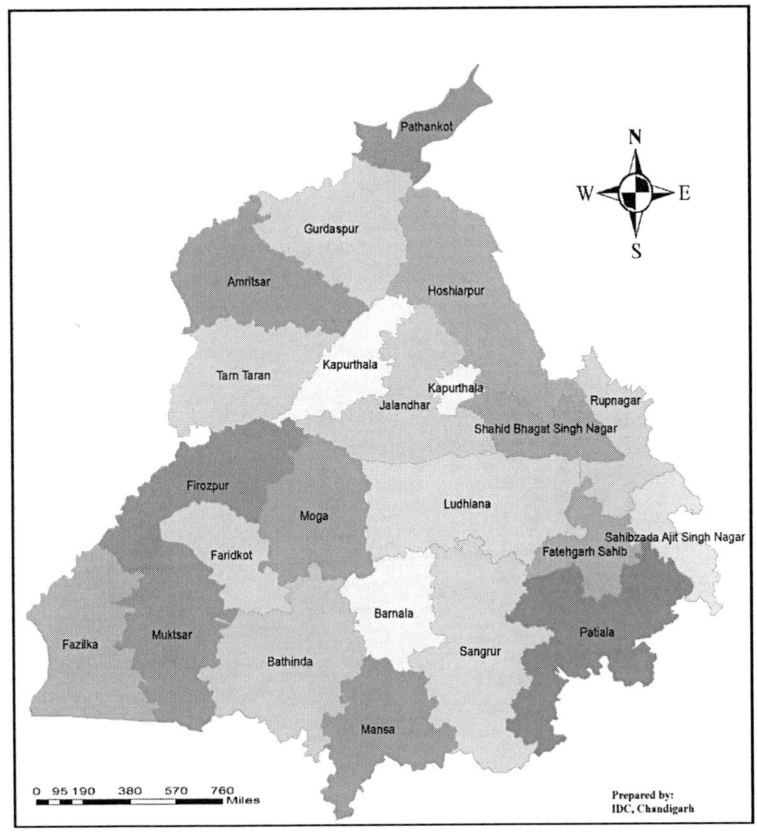

MAP 1.3

HARYANA ADMINISTRATIVE DIVISION

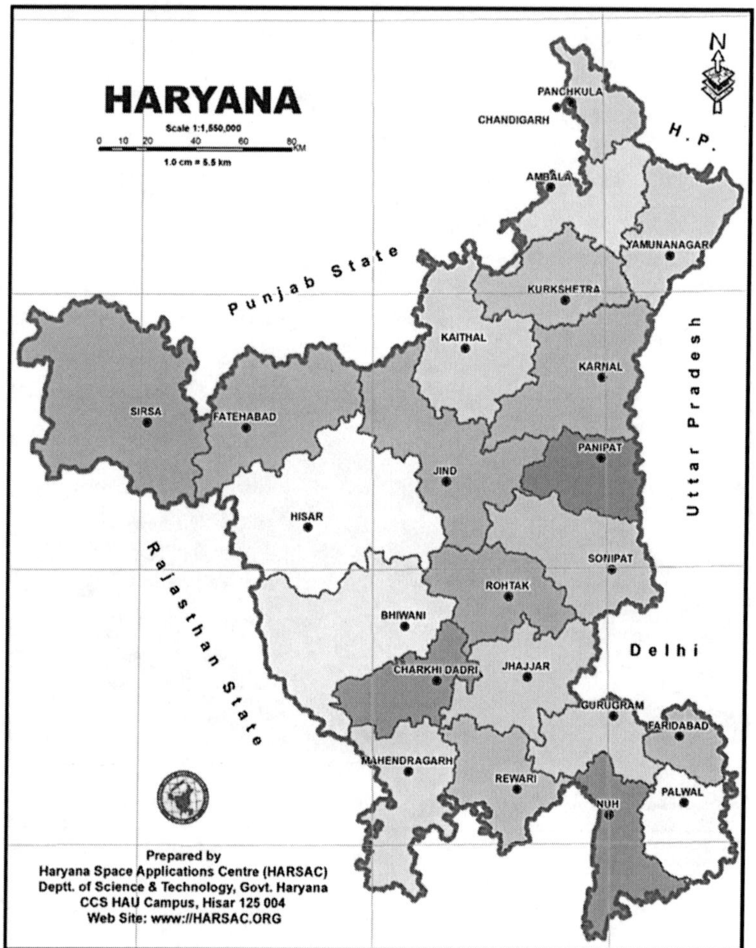

MAP 2.1

MAIN SURFACE WATER NETWORK AREA, PUNJAB

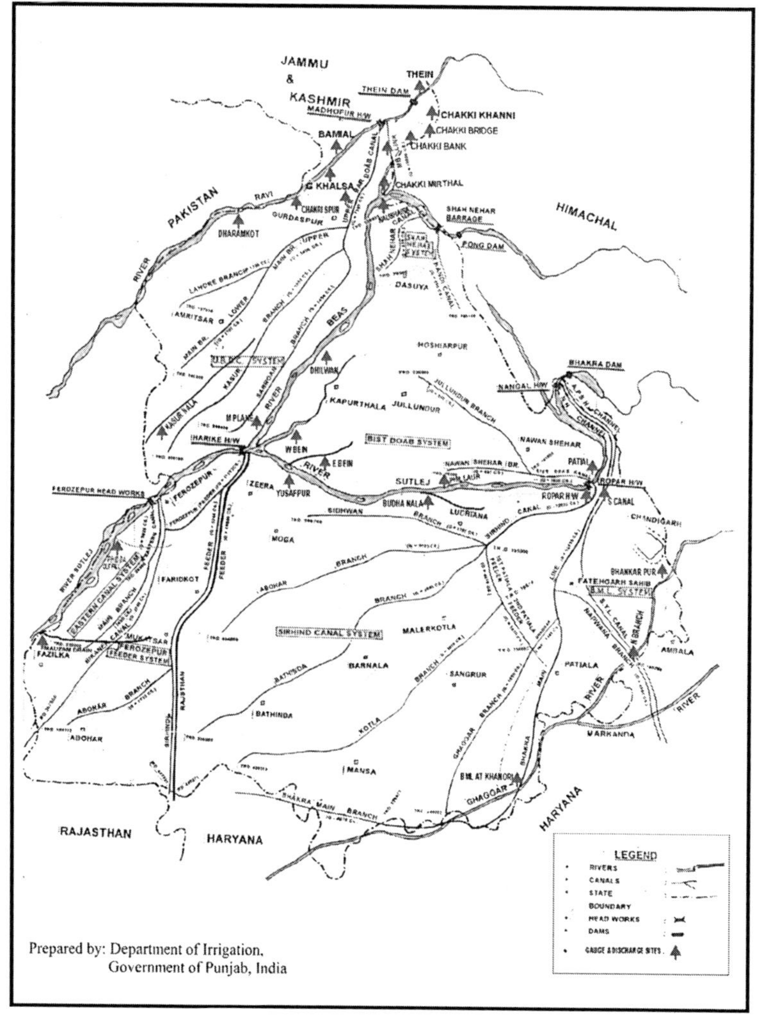

MAP 2.2

MAIN SURFACE WATER NETWORK AREA, HARYANA

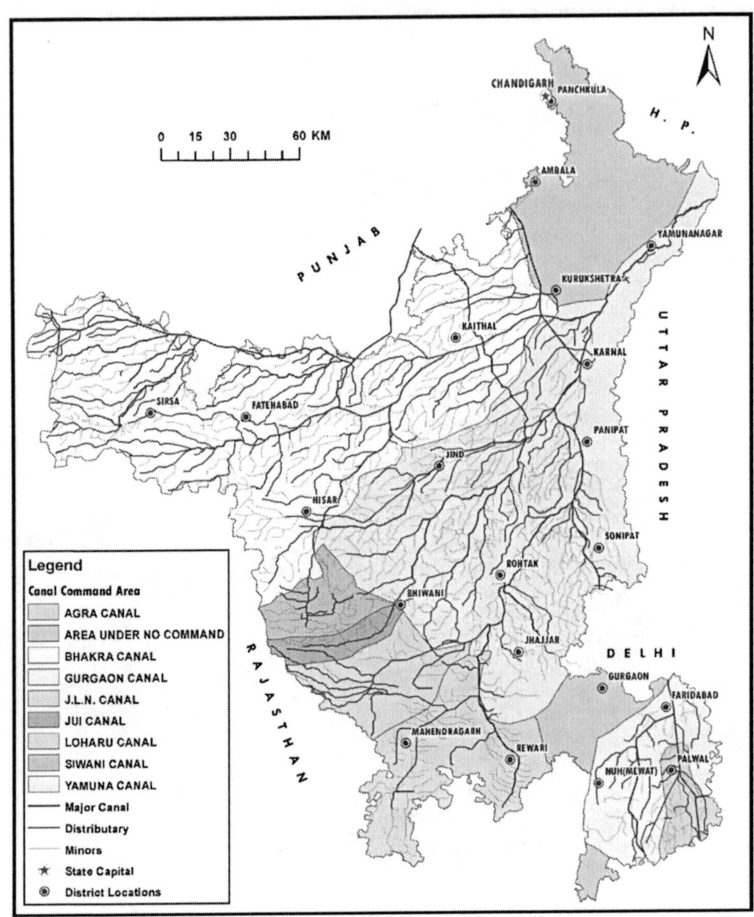

Prepared by: Deptt. of Science and Technology, Haryana.

APPENDIX B

Development Indicators: Agriculture and Socio-Economic

Appendix-1

Total Cropped Area, Area Under Cultivation of Rice and Wheat in Punjab ('000 Hectares)

Years	Total Cropped Area	Rice		Wheat	
		Area Under Rice	Percent-age of Total Cropped Area	Area Under Wheat	Percent-age of Total Cropped Area
1966-67	5171	285	5.51	1608	31.10
1967-68	5441	314	5.77	1790	32.90
1968-69	5288	345	6.52	2063	39.01
1969-70	5499	359	6.53	2166	39.39
1970-71	5678	390	6.87	2299	40.49
1971-72	5724	450	7.86	2336	40.81
1972-73	5931	476	8.03	2404	40.53
1973-74	6037	499	8.27	2338	38.73
1974-75	5904	569	9.64	2207	37.38
1975-76	6255	567	9.06	2439	38.99
1976-77	6285	680	10.82	2630	41.85
1977-78	6390	858	13.43	2617	40.95
1978-79	6630	1052	15.87	2739	41.31
1979-80	6535	1172	17.93	2813	43.05
1980-81	6763	1183	17.49	2812	41.58

Contd...

1981-82	6929	1269	18.31	2914	42.06
1982-83	6915	1322	19.12	3052	44.14
1983-84	6915	1481	21.42	3124	45.18
1984-85	7013	1644	23.44	3094	44.12
1985-86	7158	1714	23.95	3112	43.48
1986-87	7217	1786	24.75	3185	44.13
1987-88	7326	1720	23.48	3131	42.74
1988-89	7387	1778	24.07	3158	42.75
1989-90	7393	1908	25.81	3247	43.92
1990-91	7502	2015	26.86	3273	43.63
1991-92	7518	2069	27.52	3237	43.06
1992-93	7552	2072	27.44	3283	43.47
1993-94	7623	2179	28.58	3335	43.75
1994-95	7693	2265	29.44	3311	43.04
1995-96	7712	2185	28.33	3221	41.77
1996-97	7808	2159	27.65	3229	41.36
1997-98	7833	2281	29.12	3300	42.13
1998-99	7945	2519	31.71	3337	42.00
1999-2000	7847	2604	33.18	3388	43.18
2000-01	7941	2612	32.89	3408	42.92
2001-02	7941	2487	31.32	3420	43.07
2002-03	7826	2530	32.33	3375	43.13
2003-04	7907	2614	33.06	3444	43.56
2004-05	7931	2646	33.36	3481	43.89
2005-06	7868	2642	33.58	3468	44.08
2006-07	7861	2621	33.34	3467	44.10
2007-08	7870	2609	33.15	3487	44.31
2008-09	7912	2735	34.57	3526	44.57
2009-10	7876	2802	35.58	3522	44.72
2010-11	7882	2826	35.85	3510	44.53
2011-12	7902	2814	35.61	3527	44.63
2012-13	7870	2849	36.20	3517	44.69
2013-14	7848	2849	36.30	3510	44.72
2014-15	7857	2895	36.85	3505	44.61

Source: *Statistical Abstracts of Punjab* for Various Years.

Appendix-2

Production, Yield of Rice and Wheat During Different Years in Punjab

Years	Production of Total Foodgrains (000 Tonnes)	Rice		Wheat		Other Foodgrains Production (000 Tonnes)
		Production (000 Tonnes)	Yield (Kgs/Hectares)	Production (000 Tonnes)	Yield (Kgs/Hectares)	
1966-67	4178	338 (8.09)	1185	2451 (58.66)	1524	1389 (33.25)
1967-68	5368	415 (7.73)	1322	3335 (62.13)	1863	1618 (30.14)
1968-69	6212	470 (7.57)	1364	4491 (72.30)	2177	1251 (20.14)
1969-70	6924	535 (7.73)	1490	4865 (70.26)	2245	1524 (22.01)
1970-71	7305	688 (9.42)	1765	5145 (70.43)	2238	1472 (20.15)
1971-72	7900	920 (11.65)	2045	5618 (71.11)	2406	1362 (17.24)
1972-73	7677	955 (12.44)	2007	5368 (69.92)	2233	1354 (17.64)

Contd...

1973-74	7728	1140 (14.75)	2287	5181 (67.04)	2216	1407 (18.21)
1974-75	7945	1179 (14.84)	2071	5286 (66.53)	2395	1480 (18.63)
1975-76	8827	1447 (16.39)	2553	5788 (65.57)	2373	1592 (18.04)
1976-77	9199	1776 (19.31)	2611	6392 (69.49)	2430	1031 (11.21)
1977-78	10370	2497 (24.08)	2910	6642 (64.05)	2538	1231 (11.87)
1978-79	11624	3090 (26.58)	2937	7439 (64.00)	2716	1095 (9.42)
1979-80	11890	3052 (25.67)	2604	7868 (66.17)	2797	970 (8.16)
1980-81	11903	3233 (27.16)	2733	7677 (64.50)	2730	993 (8.34)
1981-82	13326	3750 (28.14)	2955	8544 (64.12)	2932	1032 (7.74)
1982-83	14146	4156 (29.38)	3144	9168 (64.81)	3004	822 (5.81)
1983-84	14781	4536 (30.69)	3063	9422 (63.74)	3015	823 (5.57)

Contd...

1984-85	16099	5052 (31.38)	3073	10176 (63.21)	3289	871 (5.41)
1985-86	17189	5485 (31.91)	3200	10988 (63.92)	3531	716 (4.17)
1986-87	16292	5949 (36.51)	3331	9447 (57.99)	2966	896 (5.5)
1987-88	17092	5442 (31.84)	3164	11084 (64.85)	3540	566 (3.31)
1988-89	17068	4925 (28.86)	2770	11580 (67.85)	3667	563 (3.3)
1989-90	18986	6697 (35.27)	3510	11666 (61.45)	3593	623 (3.28)
1990-91	19249	6506 (33.80)	3229	12159 (63.17)	3715	584 (3.03)
1991-92	19635	6739 (34.32)	3257	12309 (62.69)	3803	587 (2.99)
1992-93	20007	7026 (35.12)	3391	12399 (61.97)	3770	582 (2.91)
1993-94	21577	7645 (35.43)	3507	13378 (62.00)	4011	554 (2.57)
1994-95	21817	7662 (35.12)	3381	13542 (62.07)	4089	613 (2.81)

Contd...

1995-96	19806	6843 (34.55)	3132	12510 (63.16)	3884	453 (2.29)
1996-97	21553	7334 (34.03)	3397	13672 (63.43)	4234	547 (2.54)
1997-98	21143	7904 (37.38)	3465	12715 (60.14)	3853	524 (2.48)
1998-99	22907	7940 (34.66)	3152	14456 (63.11)	4332	511 (2.23)
1999-2000	25201	8716 (34.59)	3347	15910 (63.13)	4696	575 (2.28)
2000-01	25325	9157 (36.16)	3506	15551 (61.41)	4563	617 (2.44)
2001-02	24887	8816 (35.42)	3545	15499 (62.28)	4532	572 (2.3)
2002-03	23491	8880 (37.80)	3510	14175 (60.34)	4200	436 (1.86)
2003-04	24729	9656 (39.05)	3694	14489 (58.59)	4207	584 (2.36)
2004-05	25671	10437 (40.66)	3943	14695 (57.24)	4221	539 (2.1)
2005-06	25184	10193 (40.47)	3858	14493 (57.55)	4179	498 (1.98)

Contd...

2006-07	25313	10138 (40.05)	3868	14596 (57.66)	4210	579 (2.29)
2007-08	26815	10489 (39.12)	4019	15720 (58.62)	4507	606 (2.26)
2008-09	27326	11000 (40.25)	4022	15733 (57.58)	4462	593 (2.17)
2009-10	26947	11236 (41.70)	4010	15169 (56.29)	4307	542 (2.01)
2010-11	27846	10819 (38.85)	3828	16472 (59.15)	4693	555 (1.99)
2011-12	29085	10527 (36.19)	3741	17977 (61.81)	5097	581 (2.0)
2012-13	28551	11390 (39.89)	3998	16614 (58.19)	4724	547 (1.92)
2013-14	29443	11259 (38.24)	3952	17610 (59.81)	5017	574 (1.95)
2014-15	26708	11111 (41.60)	3838	15086 (56.48)	4304	511 (1.91)

Source: *Statistical Abstracts of Punjab* for Various Years.
Note: Figures in brackets are in percentages out of total foodgrain production.

Appendix-3

Index of Total Cropped Area, Area Under Rice Cultivation, Yield of Rice and Production of Rice in Punjab

Years	Total Cropped Area	Area	Yield	Production
1966-67	100	100	100	100
1967-68	105	110	112	123
1968-69	102	121	115	139
1969-70	106	126	126	158
1970-71	110	137	149	204
1971-72	111	158	173	272
1972-73	115	167	169	283
1973-74	117	175	193	337
1974-75	114	200	175	349
1975-76	121	199	215	428
1976-77	122	239	220	525
1977-78	124	301	246	739
1978-79	128	369	248	914
1979-80	126	411	220	903
1980-81	131	415	231	957
1981-82	134	445	249	1109
1982-83	134	464	265	1230
1983-84	134	520	258	1342
1984-85	136	577	259	1495
1985-86	138	601	270	1623
1986-87	140	627	281	1760
1987-88	142	604	267	1610
1988-89	143	624	234	1457
1989-90	143	669	296	1981
1990-91	145	707	272	1925
1991-92	145	726	275	1994

Contd..

1992-93	146	727	286	2079
1993-94	147	765	296	2262
1994-95	149	795	285	2267
1995-96	149	767	264	2025
1996-97	151	758	287	2170
1997-98	151	800	292	2338
1998-99	154	884	266	2349
1999-2000	152	914	282	2579
2000-01	154	916	296	2709
2001-02	154	873	299	2608
2002-03	151	888	296	2627
2003-04	153	917	312	2857
2004-05	153	928	333	3088
2005-06	152	927	326	3016
2006-07	152	920	326	2999
2007-08	152	915	339	3103
2008-09	153	960	339	3254
2009-10	152	983	338	3324
2010-11	152	992	323	3201
2011-12	153	987	316	3114
2012-13	152	1000	337	3370
2013-14	152	1000	334	3331
2014-15	152	1016	324	3287

Source: Appendices 1 and 2.

Appendix-4

Index of Total Cropped Area, Area Under Wheat Cultivation, Yield of Wheat and Production of Wheat in Punjab

Years	Total Cropped Area	Area	Yield	Production
1966-67	100	100	100	100
1967-68	105	111	122	136
1968-69	102	128	143	183
1969-70	106	135	147	198
1970-71	110	143	147	210
1971-72	111	145	158	229
1972-73	115	150	147	219
1973-74	117	145	145	211
1974-75	114	137	157	216
1975-76	121	152	156	236
1976-77	122	164	159	261
1977-78	124	163	167	271
1978-79	128	170	178	304
1979-80	126	175	184	321
1980-81	131	175	179	313
1981-82	134	181	192	349
1982-83	134	190	197	374
1983-84	134	194	198	384
1984-85	136	192	216	415
1985-86	138	194	232	448
1986-87	140	198	195	385
1987-88	142	195	232	452
1988-89	143	196	241	472
1989-90	143	202	236	476
1990-91	145	204	244	496
1991-92	145	201	250	502
1992-93	146	204	247	506

Contd...

1993-94	147	207	263	546
1994-95	149	206	268	553
1995-96	149	200	255	510
1996-97	151	201	278	558
1997-98	151	205	253	519
1998-99	154	208	284	590
1999-2000	152	211	308	649
2000-01	154	212	299	634
2001-02	154	213	297	632
2002-03	151	210	276	578
2003-04	153	214	276	591
2004-05	153	216	277	600
2005-06	152	216	274	591
2006-07	152	216	276	596
2007-08	152	217	296	641
2008-09	153	219	293	642
2009-10	152	219	283	619
2010-11	152	218	308	672
2011-12	153	219	334	733
2012-13	152	219	310	678
2013-14	152	218	329	718
2014-15	152	218	282	616

Source: Appendices 1 and 2.

Appendix-5

Total Cropped Area, Area Under Cultivation and Yield of Rice and Wheat in Haryana ('000 Hectares)

Years	Total Cropped Area	Rice		Wheat	
		Area Under Rice	Percent-age of Total Cropped Area	Area Under Wheat	Percent-age of Total Cropped Area
1966-67	4599	192	4.17	743	16.16
1967-68	5150	217	4.21	841	16.33
1968-69	4053	229	5.65	898	22.16
1969-70	4941	240.8	4.87	1017.3	20.59
1970-71	4957	269.2	5.43	1129.3	22.78
1971-72	5048	291	5.76	1177	23.32
1972-73	5188	291.4	5.62	1270.6	24.49
1973-74	5150	291.7	5.66	1176.5	22.84
1974-75	4842	275.4	5.69	1117.4	23.08
1975-76	5451	303.5	5.57	1226	22.49
1976-77	5282	329.6	6.24	1349	25.54
1977-78	5435	371	6.83	1359.8	25.02
1978-79	5522	458.6	8.30	1481.5	26.83
1979-80	4862	509.3	10.48	1476.8	30.37
1980-81	5462	483.9	8.86	1479	27.08
1981-82	5826	504.6	8.66	1561.9	26.81
1982-83	5306	489.5	9.23	1723.4	32.48
1983-84	5688	560.6	9.86	1793.2	31.53
1984-85	5512	557.3	10.11	1704.7	30.93
1985-86	5601	584	10.43	1701.3	30.37
1986-87	5662	628	11.09	1782.4	31.48
1987-88	4686	464.3	9.91	1731	36.94
1988-89	6012	601.7	10.01	1827	30.39
1989-90	5651	641.5	11.35	1857.2	32.86

Contd...

1990-91	5919	661.2	11.17	1850.1	31.26
1991-92	5570	637	11.44	1805.8	32.42
1992-93	5853	707.4	12.09	1963.4	33.55
1993-94	5815	755	12.98	1993.6	34.28
1994-95	5989	796.1	13.29	1985.3	33.15
1995-96	5974	830	13.89	1972.1	33.01
1996-97	6074	830.5	13.67	2017	33.21
1997-98	6143	913.7	14.87	2057	33.49
1998-99	6320	1086	17.18	2188	34.62
1999-2000	6029	1083.1	17.96	2316.5	38.42
2000-01	6115	1054.3	17.24	2354.8	38.51
2001-02	6318	1027.5	16.26	2299.9	36.40
2002-03	6035	905.7	15.01	2267.1	37.57
2003-04	6388	1015.2	15.89	2315.4	36.25
2004-05	6425	1024.2	15.94	2316.7	36.06
2005-06	6509	1046.6	16.08	2302.7	35.38
2006-07	6407	1042	16.26	2377.1	37.10
2007-08	6458	1072.5	16.61	2460.7	38.10
2008-09	6500	1211.2	18.63	2461.4	37.87
2009-10	6351	1206.4	19.00	2487.7	39.17
2010-11	6505	1243.3	19.11	2504	38.49
2011-12	6489	1234.1	19.02	2531.3	39.01
2012-13	6376	1206.3	18.92	2496.9	39.16
2013-14	6471	1244.6	19.23	2499.1	38.62
2014-15	6536	1277.9	19.55	2628.1	40.21

Source: *Statistical Abstracts of Haryana* for Various Years.

Appendix-6
Production, Yield of Rice and Wheat During Different Years in Haryana

Years	Production of Total Foodgrains (000 Tonnes)	Rice		Wheat		Other Foodgrains Production (000 Tonnes)
		Production (000 Tonnes)	Yield (Kgs/Hectares)	Production (000 Tonnes)	Yield (Kgs/Hectares)	
1966-67	2581	223 (8.64)	1161	1059 (41.03)	1425	1299 (50.33)
1967-68	3946	287 (7.27)	1324	1438 (36.44)	1710	2221 (56.28)
1968-69	2764	272 (9.84)	1186	1529 (55.32)	1703	963 (34.84)
1969-70	4626	372 (8.04)	1545	2147 (46.41)	2111	2107 (45.55)
1970-71	4732	460 (9.72)	1697	2342 (49.49)	2074	1930 (40.79)
1971-72	4539	536 (11.81)	1843	2402 (52.92)	2041	1601 (35.27)
1972-73	3948	462 (11.70)	1586	1231 (31.18)	1757	2255 (57.12)

Contd...

1973-74	3839	540 (14.07)	1851	1811 (47.17)	1539	1488 (38.76)
1974-75	3340	393 (11.77)	1425	1954 (58.50)	1748	993 (29.73)
1975-76	5050	625 (12.38)	2063	2428 (48.08)	1980	1997 (39.54)
1976-77	5259	815 (15.50)	2470	2735 (52.01)	2029	1709 (32.50)
1977-78	5341	965 (18.07)	2608	2845 (53.27)	2093	1531 (28.67)
1978-79	6358	1228 (19.31)	2680	3398 (53.44)	2293	1732 (27.24)
1979-80	5028	941 (18.72)	1852	3295 (65.53)	2231	792 (15.75)
1980-81	6045	1259 (20.83)	2606	3490 (57.73)	2360	1296 (21.44)
1981-82	6040	1252 (20.73)	2475	3686 (61.03)	2357	1102 (18.25)
1982-83	6650	1276 (19.19)	2604	4347 (65.37)	2524	1027 (15.44)
1983-84	6886	1332 (19.34)	2485	4458 (64.74)	2499	1096 (15.92)

Contd...

1984-85	6838	1363 (19.93)	2447	4421 (64.65)	2593	1054 (15.41)
1985-86	8141	1633 (20.06)	2797	5260 (64.61)	3094	1248 (15.33)
1986-87	7635	1543 (20.21)	2457	5057 (66.23)	2836	1035 (13.56)
1987-88	6302	1077 (17.09)	2321	4861 (77.13)	2808	364 (5.78)
1988-89	9502	1443 (15.19)	2397	6225 (65.51)	3407	1834 (19.30)
1989-90	8652	1750 (20.23)	2730	5907 (68.27)	3181	995 (11.50)
1990-91	9561	1834 (19.18)	2775	6436 (67.32)	3479	1291 (13.50)
1991-92	9093	1803 (19.83)	2831	6496 (71.44)	3597	794 (8.73)
1992-93	10251	1880 (18.34)	2659	7108 (69.34)	3621	1263 (12.32)
1993-94	10255	2061 (20.10)	2730	7217 (70.38)	3619	977 (9.53)
1994-95	10994	2230 (20.28)	2802	7297 (66.37)	3676	1467 (13.34)

Contd...

1995-96	10137	1847 (18.22)	1847	7291 (71.92)	3697	999 (9.85)
1996-97	11448	2463 (21.51)	2967	7826 (68.36)	3880	1159 (10.12)
1997-98	11348	2556 (22.52)	2800	7528 (66.34)	3660	1264 (11.14)
1998-99	12123	2432 (20.06)	2239	8568 (70.68)	3916	1123 (9.26)
1999-2000	13063	2583 (19.77)	2385	9650 (73.87)	4165	830 (6.35)
2000-01	13294	2695 (20.27)	2557	9699 (72.96)	4106	900 (6.77)
2001-02	13298	2726 (20.50)	2652	9437 (70.97)	4103	1135 (8.54)
2002-03	12329	2468 (20.02)	2724	9188 (74.52)	4053	673 (5.46)
2003-04	13193	2790 (21.15)	2749	9114 (69.08)	3937	1289 (9.77)
2004-05	13109	3010 (22.96)	2939	9043 (68.98)	3901	1056 (8.06)
2005-06	12998	3194 (24.57)	3051	8853 (68.11)	3844	951 (7.32)

Contd...

2006-07	14763	3375 (22.86)	3238	10059 (68.14)	4232	1329 (9.00)
2007-08	15308	3606 (23.56)	3361	10232 (66.84)	4158	1470 (9.60)
2008-09	15614	3299 (21.13)	2724	11360 (72.76)	4614	955 (6.12)
2009-10	15774	3628 (23.00)	3008	10488 (66.49)	4215	1658 (10.51)
2010-11	16041	3465 (21.60)	2788	11578 (72.18)	4624	998 (6.22)
2011-12	17957	3757 (20.92)	3044	13119 (73.06)	5183	1081 (6.02)
2012-13	16226	3941 (24.29)	3268	11117 (68.51)	4452	1168 (7.20)
2013-14	16974	4041 (23.81)	3248	11800 (69.52)	4722	1133 (6.67)
2014-15	16748	4007 (23.93)	3113	10707 (63.93)	3981	2034 (12.14)

Source: *Statistical Abstracts of Haryana* for Various Years.

Appendix-7

Index of Total Cropped Area, Area Under Wheat Cultivation, Yield of Wheat and Production of Wheat in Haryana

Years	Total Cropped Area	Area	Yield	Production
1966-67	100	100	100	100
1967-68	112	113	120	136
1968-69	88	121	120	144
1969-70	107	137	148	203
1970-71	108	152	146	221
1971-72	110	158	143	227
1972-73	113	171	123	116
1973-74	112	158	108	171
1974-75	105	150	123	185
1975-76	119	165	139	229
1976-77	115	182	142	258
1977-78	118	183	147	269
1978-79	120	199	161	321
1979-80	106	199	157	311
1980-81	119	199	166	330
1981-82	127	210	165	348
1982-83	115	232	177	410
1983-84	124	241	175	421
1984-85	120	229	182	417
1985-86	122	229	217	497
1986-87	123	240	199	478
1987-88	102	233	197	459
1988-89	131	246	239	588
1989-90	123	250	223	558
1990-91	129	249	244	608
1991-92	121	243	252	613

Contd...

1992-93	127	264	254	671
1993-94	126	268	254	681
1994-95	130	267	258	689
1995-96	130	265	259	688
1996-97	132	271	272	739
1997-98	134	277	257	711
1998-99	137	294	275	809
1999-2000	131	312	292	911
2000-01	133	317	288	916
2001-02	137	310	288	891
2002-03	131	305	284	868
2003-04	139	312	276	861
2004-05	140	312	274	854
2005-06	142	310	270	836
2006-07	139	320	297	950
2007-08	140	331	292	966
2008-09	141	331	324	1073
2009-10	138	335	296	990
2010-11	141	337	324	1093
2011-12	141	341	364	1239
2012-13	139	336	312	1050
2013-14	141	336	331	1114
2014-15	142	354	279	1011

Source: Appendices 5 and 6.

Appendix-8

Index of Total Cropped Area, Area Under Rice Cultivation, Yield of Rice and Production of Rice in Haryana

Years	Total Cropped Area	Area	Yield	Production
1966-67	100	100	100	100
1967-68	112	88	114	129
1968-69	88	119	102	122
1969-70	107	125	133	167
1970-71	108	140	146	206
1971-72	110	152	159	240
1972-73	113	152	137	207
1973-74	112	152	159	242
1974-75	105	143	123	176
1975-76	119	158	178	280
1976-77	115	172	213	365
1977-78	118	193	225	433
1978-79	120	239	231	551
1979-80	106	265	160	422
1980-81	119	252	224	565
1981-82	127	263	213	561
1982-83	115	255	224	572
1983-84	124	292	214	597
1984-85	120	290	211	611
1985-86	122	304	241	732
1986-87	123	327	212	692
1987-88	102	242	200	483
1988-89	131	313	206	647
1989-90	123	334	235	785
1990-91	129	344	239	822
1991-92	121	332	244	809
1992-93	127	368	229	843

Contd...

1993-94	126	393	235	924
1994-95	130	415	241	1000
1995-96	130	432	159	828
1996-97	132	433	256	1104
1997-98	134	476	241	1146
1998-99	137	566	193	1091
1999-2000	131	564	205	1158
2000-01	133	549	220	1209
2001-02	137	535	228	1222
2002-03	131	472	235	1107
2003-04	139	529	237	1251
2004-05	140	533	253	1350
2005-06	142	545	263	1432
2006-07	139	543	279	1513
2007-08	140	559	289	1617
2008-09	141	631	235	1479
2009-10	138	628	259	1627
2010-11	141	648	240	1554
2011-12	141	643	262	1685
2012-13	139	628	281	1767
2013-14	141	648	280	1812
2014-15	142	666	268	1797

Source: Appendices 5 and 6.

Appendix-9

Total Cropped Area, Area Under Cultivation of Main Non-Foodgrain Crops (Cotton, Sugarcane and Oilseeds) in Punjab ('000 Hectares)

Years	Total Cropped Area	Sugarcane		Cotton		Total Oilseeds	
		Area Under Sugarcane	Percent of Total Cropped Area	Area Under Cotton	Percent of Total Cropped Area	Area Under Oil Seeds	Percent of Total Cropped Area
1966-67	5171	156	3.02	435	8.42	323	6.25
1967-68	5441	137	2.52	419	7.69	399	7.33
1968-69	5288	157	2.97	392	7.41	307	5.81
1969-70	5499	149	2.71	408	7.41	294	5.35
1970-71	5678	128	2.25	397	7.00	296	5.21
1971-72	5724	103	1.80	475	8.30	319	5.57
1972-73	5931	102	1.72	445	7.50	351	5.92
1973-74	6037	110	1.82	520	8.61	357	5.91
1974-75	5904	123	2.08	543	9.19	372	6.30
1975-76	6255	114	1.82	579	9.25	315	5.04

Contd...

1976-77	6285	113	1.80	551	8.77	250	3.98
1977-78	6390	115	1.80	608	9.51	291	4.55
1978-79	6630	108	1.63	683	10.29	230	3.47
1979-80	6535	77	1.18	628	9.61	199	3.05
1980-81	6763	71	1.05	641	9.48	238	3.52
1981-82	6929	104	1.50	685	9.88	225	3.25
1982-83	6915	104	1.50	721	10.43	187	2.70
1983-84	6915	84	1.21	648	9.37	156	2.26
1984-85	7013	79	1.13	468	6.67	198	2.82
1985-86	7158	78	1.09	555	7.76	211	2.95
1986-87	7217	97	1.34	563	7.80	160	2.22
1987-88	7326	106	1.45	618	8.44	155	2.12
1988-89	7387	97	1.31	758	10.25	202	2.73
1989-90	7393	103	1.39	649	8.78	120	1.62
1990-91	7502	101	1.35	756	10.07	104	1.39
1991-92	7518	109	1.45	696	9.26	213	2.83
1992-93	7552	112	1.48	700	9.27	194	2.57

Contd...

1993-94	7623	77	1.01	578	7.58	150	1.97
1994-95	7693	80	1.04	604	7.85	153	1.99
1995-96	7712	136	1.76	666	8.64	237	3.07
1996-97	7808	173	2.22	716	9.17	246	3.15
1997-98	7833	126	1.61	724	9.25	139	1.77
1998-99	7945	103	1.30	562	7.08	160	2.01
1999-2000	7847	108	1.38	476	6.07	98	1.25
2000-01	7941	121	1.52	473	5.96	86	1.08
2001-02	7941	142	1.79	607	7.64	83	1.05
2002-03	7826	153	1.96	449	5.73	98	1.25
2003-04	7907	123	1.56	449	5.68	86	1.09
2004-05	7931	85	1.07	507	6.39	92	1.16
2005-06	7868	85	1.08	554	7.04	82	1.04
2006-07	7861	99	1.26	607	7.72	70	0.89
2007-08	7870	108	1.37	600	7.62	59	0.75
2008-09	7912	81	1.02	527	6.66	60	0.76

Contd...

2009-10	7876	60	0.76	511	6.49	62	0.79
2010-11	7882	70	0.89	483	6.13	56	0.71
2011-12	7902	80	1.01	515	6.52	52	0.66
2012-13	7870	82	1.04	481	6.11	51	0.65
2013-14	7848	89	1.13	445	5.67	47	0.60
2014-15	7857	94	1.20	420	5.35	46	0.59

Source: *Statistical Abstracts of Punjab.*

Appendix-10

Production and Yield of Main Non-Foodgrain Crops (Cotton, Sugarcane and Oilseeds) in Punjab ('000 Tonnes)

Years	Production of Total Non-Foodgrains	Sugarcane			Cotton			Total Oilseeds		
		Production	Percentage Out of Total Non-Foodgrains	Yield (Kgs/ Hectares)	Production	Percentage Out of Total Non-Foodgrains	Yield (Kgs/ Hectares)	Production	Percentage Out of Total Non-Foodgrains	Yield (Kgs/ Hectares)
1966-67	820	436	53.17	2795	132	16.12	304	252	30.73	780
1967-68	934	480	51.39	3504	140	14.99	335	314	33.62	787
1968-69	900	516	57.33	3287	134	14.93	343	249	27.67	811
1969-70	976	618	63.32	4148	144	14.75	353	213	21.82	724
1970-71	1578	527	33.40	4117	147	9.34	371	233	14.77	787
1971-72	1647	403	24.47	3913	173	10.49	364	272	16.51	853
1972-73	1753	469	26.75	4598	180	10.28	405	268	15.29	764
1973-74	1969	582	29.56	5291	196	9.97	378	295	14.98	826
1974-75	2106	615	29.20	5000	203	9.63	374	291	13.82	782
1975-76	2111	613	29.04	5377	210	9.95	363	263	12.46	835
1976-77	1933	607	31.40	5372	193	9.99	350	206	10.66	824
1977-78	2101	652	31.03	5670	209	9.96	344	214	10.19	735
1978-79	2120	612	28.87	5667	224	10.58	328	178	8.40	774

Contd...

1979-80	1766	393	22.25	5104	215	12.18	342	157	8.89	789
1980-81	5285	392	7.42	5521	200	3.79	312	187	3.54	786
1981-82	7559	601	7.95	5779	217	2.87	317	173	2.29	769
1982-83	7691	634	8.24	6096	207	2.70	287	140	1.82	749
1983-84	6357	553	8.70	6583	120	1.89	186	119	1.87	763
1984-85	6360	492	7.74	6228	210	3.31	450	199	3.13	1005
1985-86	6651	504	7.58	6462	238	3.58	428	202	3.04	957
1986-87	7967	611	7.67	6299	288	3.61	511	140	1.76	875
1987-88	7915	582	7.35	5491	316	3.99	511	188	2.38	1213
1988-89	8263	600	7.26	6186	360	4.36	475	145	1.75	718
1989-90	9086	650	7.15	6311	417	4.59	643	107	1.18	892
1990-91	8020	601	7.49	5950	351	4.38	465	93	1.16	894
1991-92	9536	693	7.27	6358	410	4.30	589	267	2.80	1254
1992-93	8914	688	7.72	6143	399	4.48	570	236	2.65	1216
1993-94	6459	468	7.25	6078	258	4.00	447	180	2.79	1200
1994-95	7203	495	6.87	6188	303	4.21	502	188	2.61	1229
1995-96	10876	888	8.16	6529	289	2.66	434	288	2.65	1215
1996-97	13249	1022	7.71	5908	317	2.40	443	336	2.54	1366
1997-98	8306	715	8.61	5675	159	1.92	220	145	1.75	1043

Contd...

1998-99	6896	613	8.89	5951	101	1.47	180	167	2.42	1044
1999-2000	7834	676	8.63	6259	162	2.06	339	104	1.33	1061
2000-01	9056	777	8.58	6421	203	2.24	429	88	0.97	1023
2001-02	10640	925	8.69	6514	222	2.09	366	84	0.79	1012
2002-03	10464	902	8.62	5895	184	1.76	411	85	0.81	867
2003-04	8202	662	8.07	5382	251	3.06	560	103	1.26	1198
2004-05	7357	511	6.95	6012	355	4.82	700	104	1.41	1130
2005-06	7345	491	6.69	5776	407	5.55	735	91	1.24	1110
2006-07	8976	601	6.70	6071	455	5.07	750	78	0.87	1114
2007-08	9122	657	7.20	6083	400	4.39	667	77	0.84	1305
2008-09	7031	467	6.64	5765	388	5.52	736	73	1.04	1217
2009-10	5801	370	6.38	6167	341	5.88	667	84	1.45	1355
2010-11	6341	417	6.58	5957	310	4.89	642	73	1.15	1304
2011-12	7038	467	6.64	5838	276	3.92	536	69	0.98	1327
2012-13	7988	483	6.05	5890	277	3.47	576	70	0.88	1373
2013-14	8707	552	6.34	6202	254	2.92	571	60	0.69	1277
2014-15	8718	581	6.66	6181	228	2.62	543	58	0.67	1261

Source: *Statistical Abstracts of Punjab.*

Appendix-11

Total Cropped Area, Area Under Cultivation of Main Non-Foodgrain Crops (Cotton, Sugarcane and Oilseeds) in Haryana ('000 Hectares)

Years	Total Cropped Area	Sugarcane		Cotton		Total Oilseeds	
		Area	Percent of Total Cropped Area	Area	Percent of Total Cropped Area	Area	Percent of Total Cropped Area
1966-67	4599	150	3.26	183	3.98	212	4.61
1967-68	5150	121	2.35	241	4.68	263	5.11
1968-69	4053	160	3.95	212	5.23	84	2.07
1969-70	4941	169	3.42	194	3.93	135	2.73
1970-71	4957	156	3.15	193	3.89	143	2.88
1971-72	5048	114	2.26	242	4.79	177	3.51
1972-73	5188	126	2.43	258	4.97	222	4.28
1973-74	5150	150	2.91	251	4.87	180	3.50
1974-75	4842	161	3.33	247	5.10	214	4.42
1975-76	5451	158	2.90	255	4.68	154	2.83
1976-77	5282	169	3.20	244	4.62	118	2.23

Contd...

1977-78	5435	196	3.61	1829	33.65	189	3.48
1978-79	5522	190	3.44	285	5.16	131	2.37
1979-80	4862	127	2.61	317	6.52	221	4.55*
1980-81	5462	113	2.07	316	5.79	311	5.69
1981-82	5826	147	2.52	1530	26.26	293	5.03*
1982-83	5306	147	2.77	397	7.48	276	5.20
1983-84	5688	134	2.36	410	7.21	205	3.60
1984-85	5512	136	2.47	493	8.94	292	5.30*
1985-86	5601	104	1.86	344	6.14	380	6.78
1986-87	5662	126	2.23	381	6.73	289	5.10
1987-88	4686	142	3.03	416	8.88	336	7.17
1988-89	6012	131	2.18	433	7.20	391	6.50
1989-90	5651	137	2.42	472	8.35	447	7.91
1990-91	5919	148	2.50	491	8.30	489	8.26
1991-92	5570	162	2.91	506	9.08	701	12.59
1992-93	5853	138	2.36	533	9.11	589	10.06
1993-94	5815	112	1.93	563	9.68	595	10.23

Contd...

1994-95	5989	119	1.99	557	9.30	619	10.34
1995-96	5974	144	2.41	652	10.91	611	10.23
1996-97	6074	162	2.67	653	10.75	673	11.08
1997-98	6143	142	2.31	632	10.29	616	10.03
1998-99	6320	128	2.03	583	9.22	526	8.32
1999-2000	6029	137	2.27	544	9.02	463	7.68
2000-01	6115	143	2.34	555	9.08	414	6.77
2001-02	6318	161	2.55	629	9.96	545	8.63
2002-03	6035	189	3.13	518	8.58	621	10.29
2003-04	6388	160	2.50	526	8.23	633	9.91
2004-05	6425	133	2.07	621	9.67	715	11.13
2005-06	6509	129	1.98	584	8.97	736	11.31
2006-07	6407	140	2.19	530	8.27	621	9.69
2007-08	6458	140	2.17	483	7.48	529	8.19
2008-09	6500	90	1.38	455	7.00	541	8.32
2009-10	6351	74	1.17	507	7.98	531	8.36*
2010-11	6505	85	1.31	493	7.58	521	8.01

Contd...

2011-12	6489	95	1.46	602	9.28	546	8.41
2012-13	6376	101	1.58	593	9.30	580	9.10
2013-14	6471	101	1.56	568	8.78	549	8.48
2014-15	6536	96	1.47	647	9.90	495	7.57

Source: *Statistical Abstracts of Haryana* for Various Years.

Appendix-12

Production and Yield of Main Non-Foodgrain Crops (Cotton, Sugarcane and Oilseeds) in Haryana ('000 Tonnes)

Years	Production of Total Non-Foodgrains	Sugarcane			Cotton			Total Oilseeds		
		Production	Percentage Out of Total Non-Foodgrains	Yield (Kgs/ Hectares)	Production	Percentage Out of Total Non-Foodgrains	Yield (Kgs/ Hectares)	Production	Percentage Out of Total Non-Foodgrains	Yield (Kgs/ Hectares)
1966-67	890	510	57.30	3400	288	32.36	1574	92	10.34	434
1967-68	964	471	48.86	3893	374	38.80	1552	121	12.55	460
1968-69	1049	669	63.78	4181	337	32.13	1590	43	4.10	512
1969-70	1220	792	64.92	4686	340	27.87	1753	89	7.30	659
1970-71	1145	707	61.75	4532	353	30.83	1829	99	8.65	692
1971-72	1054	514	48.77	4509	439	41.65	1814	100	9.49	565
1972-73	1143	560	48.98	4444	423	36.99	1640	107	9.36	482
1973-74	1105	593	53.66	3953	441	39.90	1757	61	5.52	339
1974-75	1191	591	49.61	3671	426	35.76	1725	149	12.51	696
1975-76	1232	687	55.76	4348	465	37.74	1824	79	6.41	513
1976-77	1284	728	56.72	4308	478	37.24	1959	79	6.15	669
1977-78	1459	897	61.48	4577	464	31.80	254	98	6.72	519
1978-79	1381	685	49.60	3605	601	43.52	2109	95	6.88	725

Contd...

1979-80	1057	395	37.39	3110	590	55.84	1861	255	24.14	1154*
1980-81	5499	460	8.37	4071	643	11.69	2035	188	3.42	605
1981-82	6671	584	8.75	3973	680	10.19	444	298	4.47	1017*
1982-83	6457	550	8.52	3741	840	13.01	2116	147	2.28	533
1983-84	6670	593	8.89	4425	575	8.62	1402	164	2.46	800
1984-85	**6106**	**698**	11.43	**5132**	**660**	10.81	**1339**	**361**	5.91	**1236***
1985-86	6177	501	8.11	4817	745	12.06	2166	288	4.66	758
1986-87	7867	684	8.69	5429	903	11.48	2370	226	2.87	782
1987-88	6263	524	8.37	3690	690	11.02	1659	334	5.33	994
1988-89	7909	658	8.32	5023	846	10.70	1954	484	6.12	1238
1989-90	8369	736	8.79	5372	1191	14.23	2523	435	5.20	973
1990-91	9601	780	8.12	5270	1155	12.03	2352	638	6.65	1305
1991-92	11177	905	8.10	5586	1341	12.00	2650	758	6.78	1081
1992-93	8544	672	7.86	4870	1411	16.51	2647	559	6.54	949
1993-94	8414	646	7.68	5768	1124	13.36	1996	823	9.78	1383
1994-95	9246	696	7.53	5849	1371	14.83	2461	861	9.31	1391
1995-96	10156	809	7.97	5618	1284	12.64	1969	783	7.71	1282
1996-97	11532	902	7.82	5568	1507	13.07	2308	985	8.54	1464
1997-98	9102	750	8.24	5282	1107	12.16	1752	456	5.01	740

Contd...

1998-99	8467	701	8.28	5477	874	10.32	1499	653	7.71	1241
1999-2000	9551	764	8.00	5577	1304	13.65	2397	605	6.33	1307
2000-01	10124	817	8.07	5713	1383	13.66	2492	563	5.56	1360
2001-02	10799	927	8.58	5758	722	6.69	1148	805	7.45	1477
2002-03	12400	1065	8.59	5635	1038	8.37	2004	706	5.69	1137
2003-04	11683	928	7.94	5800	1407	12.04	2675	977	8.36	1543
2004-05	10976	823	7.50	6188	2075	18.91	3341	836	7.62	1169
2005-06	10504	831	7.91	6442	1502	14.30	2572	822	7.83	1117
2006-07	12229	958	7.83	6843	1814	14.83	3423	835	6.83	1345
2007-08	11388	886	7.78	6329	1885	16.55	3903	643	5.65	1216
2008-09	7921	513	6.48	5700	1858	23.46	4084	933	11.78	1725
2009-10	7817	534	6.83	7216	1926	24.64	3799	890	11.39	1676*
2010-11	8701	604	6.94	7106	1747	20.08	3544	965	11.09	1852
2011-12	10494	695	6.62	7316	2621	24.98	4354	755	7.19	1383
2012-13	10930	744	6.81	7366	2378	21.76	4010	993	9.09	1712
2013-14	10700	773	7.22	7653	2025	18.93	3565	889	8.31	1619
2014-15	10673	710	6.65	7396	1943	18.20	3003	740	6.93	1495

Source: *Statistical Abstracts of Haryana* for Various Years.

Appendix-13
Area Under Main Non-Foodgrain Crops in Punjab and Haryana ('000 Hectares)

Years	Punjab			Haryana		
	Total Cropped Area	Area Under Main Non-Foodgrains	Percentage Out of Total Cropped Area	Total Cropped Area	Area Under Main Non-Foodgrains	Percentage Out of Total Cropped Area
1966-67	5171	914	17.68	4599	545	11.85
1967-68	5441	955	17.54	5150	625	12.14
1968-69	5288	856	16.19	4053	456	11.25
1969-70	5499	851	15.47	4941	498	10.08
1970-71	5678	821	14.46	4957	492	9.93
1971-72	5724	897	15.67	5048	533	10.56
1972-73	5931	898	15.14	5188	606	11.68
1973-74	6037	987	16.34	5150	581	11.28
1974-75	5904	1038	17.57	4842	622	12.85
1975-76	6255	1008	16.11	5451	567	10.40
1976-77	6285	914	14.55	5282	531	10.05
1977-78	6390	1014	15.87	5435	2214	40.74
1978-79	6630	1021	15.39	5522	606	10.97

Contd...

1979-80	6535	904	13.83	4862	665	13.68
1980-81	6763	950	14.05	5462	740	13.55
1981-82	6929	1014	14.63	5826	1970	33.81
1982-83	6915	1012	14.64	5306	820	15.45
1983-84	6915	888	12.84	5688	749	13.17
1984-85	7013	745	10.62	5512	921	16.71
1985-86	7158	844	11.80	5601	828	14.78
1986-87	7217	820	11.36	5662	796	14.06
1987-88	7326	879	12.00	4686	894	19.08
1988-89	7387	1057	14.30	6012	955	15.88
1989-90	7393	872	11.79	5651	1056	18.69
1990-91	7502	961	12.81	5919	1128	19.06
1991-92	7518	1018	13.54	5570	1369	24.58
1992-93	7552	1006	13.33	5853	1260	21.53
1993-94	7623	805	10.56	5815	1270	21.84
1994-95	7693	837	10.88	5989	1295	21.62
1995-96	7712	1039	13.47	5974	1407	23.55
1996-97	7808	1135	14.53	6074	1488	24.50

Contd...

1997-98	7833	989	12.63	6143	1390	22.63
1998-99	7945	825	10.39	6320	1237	19.57
1999-2000	7847	682	8.69	6029	1144	18.97
2000-01	7941	680	8.57	6115	1112	18.18
2001-02	7941	832	10.47	6318	1335	21.13
2002-03	7826	700	8.94	6035	1328	22.00
2003-04	7907	658	8.32	6388	1319	20.65
2004-05	7931	684	8.62	6425	1469	22.86
2005-06	7868	721	9.16	6509	1449	22.26
2006-07	7861	776	9.87	6407	1291	20.15
2007-08	7870	767	9.75	6458	1152	17.84
2008-09	7912	668	8.44	6500	1086	16.71
2009-10	7876	633	8.04	6351	1112	17.51
2010-11	7882	609	7.73	6505	1099	16.89
2011-12	7902	647	8.19	6489	1243	19.16
2012-13	7870	614	7.80	6376	1274	19.98
2013-14	7848	581	7.40	6471	1218	18.82
2014-15	7857	560	7.13	6536	1238	18.94

Source: *Statistical Abstracts of Haryana* for Various Years.

Appendix-14

Growth in Total Cropped Area and Total Foodgrain Production in Punjab (Regression Analysis)

Equation: In (Y) = ∝ + β.T

Dependent Variables (Y)	Intercept (∝)				Slope (β)				R^2			
	1966-67 to 1980-81	1981-82 to 1995-96	1996-97 to 2014-15	1966-67 to 2014-15	1966-67 to 1980-81	1981-82 to 1995-96	1996-97 to 2014-15	1966-67 to 2014-15	1966-67 to 1980-81	1981-82 to 1995-96	1996-97 to 2014-15	1966-67 to 2014-15
Total Cropped Area	8.54	8.83	8.97	8.66	0.02 (19.80)***	0.009 (20.27)***	(-) 0.00001 (0.07)	0.008 (15.53)***	0.97	0.97	0.00	0.84
Production of Foodgrains	8.48	9.52	10.03	8.81	0.06 (12.30)***	0.03 (11.53)***	0.01 (8.18)***	0.03 (21.33)***	0.92	0.91	0.80	0.90

Note: (i) Figures in brackets are t values.
(ii) t values are significant at: ***1 percent.

Appendix-15
Growth in Total Cropped Area and Total Foodgrain Production in Haryana

(Regression Analysis)
Equation: In (Y) = ∝ + β.T

Dependent Variables (Y)	Intercept (∝)				Slope (β)				R^2			
	1966-67 to 1980-81	1981-82 to 1995-96	1996-97 to 2014-15	1966-67 to 2014-15	1966-67 to 1980-81	1981-82 to 1995-96	1996-97 to 2014-15	1966-67 to 2014-15	1966-67 to 1980-81	1981-82 to 1995-96	1996-97 to 2014-15	1966-67 to 2014-15
Total Cropped Area	8.44	8.60	8.72	8.48	0.01 (2.77)***	0.005 (1.42)	0.003 (4.97)***	0.007 (13.56)***	0.37	0.14	0.59	0.80
Production of Foodgrains	8.02	8.14	8.66	8.13	0.04 (4.21)***	0.04 (7.26)***	0.02 (8.10)***	0.04 (28.13)***	0.58	0.80	0.82	0.94

Note: (i) Figures in brackets are t values.
(ii) t values are significant at: **5 per cent and ***1 per cent.

Appendix-16
Growth in Area, Production and Yield of Wheat in Punjab

(Regression Analysis)
Equation: In (Y) = ∝ + β.T

Dependent Variables (Y)	Intercept (∝)				Slope (β)				R^2			
	1966-67 to 1980-81	1981-82 to 1995-96	1996-97 to 2014-15	1966-67 to 2014-15	1966-67 to 1980-81	1981-82 to 1995-96	1996-97 to 2014-15	1966-67 to 2014-15	1966-67 to 1980-81	1981-82 to 1995-96	1996-97 to 2014-15	1966-67 to 2014-15
Area	7.49	7.90	8.10	7.70	0.03 (8.38)***	0.007 (6.67)***	0.004 (8.10)***	0.01 (12.74)***	0.84	0.77	0.79	0.78
Production	8.08	8.62	9.53	8.43	0.06 (7.70)***	0.03 (9.76)***	0.01 (3.92)***	0.03 (17.58)***	0.82	0.88	0.48	0.87
Yield	7.50	7.63	8.33	7.63	0.03 (6.37)***	0.02 (7.94)***	0.006 (2.50)**	0.01 (20.09)***	0.76	0.83	0.27	0.90

Note: (i) Figures in brackets are t values.
(ii) t values are significant at: ***1 per cent and **5 per cent.

Appendix-17
Growth in Area, Production and Yield of Rice in Punjab

(Regression Analysis)
Equation: In (Y) = ∝ + β.T

Dependent Variables (Y)	Intercept (∝)				Slope (β)				R^2			
	1966-67 to 1980-81	1981-82 to 1995-96	1996-97 to 2014-15	1966-67 to 2014-15	1966-67 to 1980-81	1981-82 to 1995-96	1996-97 to 2014-15	1966-67 to 2014-15	1966-67 to 1980-81	1981-82 to 1995-96	1996-97 to 2014-15	1966-67 to 2014-15
Area	5.47	7.19	7.76	6.11	0.10 (19.27)***	0.04 (12.61)***	0.01 (8.09)***	0.05 (16.03)***	0.97	0.93	0.79	0.85
Production	5.67	8.29	8.96	6.73	0.17 (30.96)***	0.05 (8.62)***	0.02 (10.84)***	0.07 (15.29)***	0.99	0.85	0.87	0.83
Yield	1.17	8.01	8.14	7.52	0.06 (10.95)***	0.007 (2.10)**	0.01 (5.67)***	0.02 (11.73)***	0.90	0.25	0.65	0.75

Note: (i) Figures in brackets are t values.
(ii) t values are significant at: ***1 per cent and **5 per cent.

Appendix-18
Growth in Area, Production and Yield of Wheat in Haryana

(Regression Analysis)
Equation: In (Y) = ∝ + β.T

Dependent Variables (Y)	Intercept (∝)				Slope (β)				R^2			
	1966-67 to 1980-81	1981-82 to 1995-96	1996-97 to 2014-15	1966-67 to 2014-15	1966-67 to 1980-81	1981-82 to 1995-96	1996-97 to 2014-15	1966-67 to 2014-15	1966-67 to 1980-81	1981-82 to 1995-96	1996-97 to 2014-15	1966-67 to 2014-15
Area	6.70	7.18	7.65	6.93	0.04 (9.45)***	0.01 (7.48)***	0.01 (9.40)***	0.02 (21.59)***	0.87	0.81	0.84	0.91
Production	7.40	7.52	8.33	7.40	0.05 (23.0)***	0.05 (13.47)***	0.02 (6.86)***	0.05 (23.0)***	0.92	0.93	0.73	0.92
Yield	6.53	7.25	7.93	7.42	0.005 (50.52)***	0.03 (10.48)***	0.01 (3.66)***	0.02 (21.61)***	0.99	0.89	0.44	0.91

Note: (i) Figures in brackets are t values.
(ii) t values are significant at: ***1 per cent.

Appendix-19

Growth in Area, Production and Yield of Rice in Haryana

(Regression Analysis)
Equation: In (Y) = $\propto + \beta.T$

Dependent Variables (Y)	Intercept (∝)				Slope (β)				R^2			
	1966-67 to 1980-81	1981-82 to 1995-96	1996-97 to 2014-15	1966-67 to 2014-15	1966-67 to 1980-81	1981-82 to 1995-96	1996-97 to 2014-15	1966-67 to 2014-15	1966-67 to 1980-81	1981-82 to 1995-96	1996-97 to 2014-15	1966-67 to 2014-15
Area	5.22	5.65	6.26	5.46	0.06 (11.73)***	0.03 (6.72)***	0.01 (6.47)***	0.04 (29.0)***	0.91	0.78	0.71	0.95
Production	5.37	6.50	6.81	5.97	0.12 (10.96)***	0.04 (5.21)***	0.03 (14.59)***	0.05 (21.89)***	0.90	0.68	0.93	0.88
Yield	7.06	7.87	7.46	7.41	0.05 (5.95)***	(-)0.003 (0.19)	0.01 (3.52)***	0.01 (8.96)***	0.73	0.00	0.42	0.63

Note: (i) Figures in brackets are t values.
(ii) t values are significant at: ***1 per cent.

Appendix-20
Growth in Area, Production of Non-Foodgrains in Punjab

Dependent Variables (Y)	Intercept (∝)				Slope (β)				R^2			
	1966-67 to 1980-81	1981-82 to 1995-96	1996-97 to 2014-15	1966-67 to 2014-15	1966-67 to 1980-81	1981-82 to 1995-96	1996-97 to 2014-15	1966-67 to 2014-15	1966-67 to 1980-81	1981-82 to 1995-96	1996-97 to 2014-15	1966-67 to 2014-15
Area	6.78	6.80	6.82	6.95	0.007 (2.03)**	0.002 (0.32)	-0.02 (5.44)**	-0.008 (6.73)**	0.24	0.008	0.64	0.49
Production	6.69	8.83	9.15	7.41	0.09 (6.17)***	0.02 (2.01)**	-0.01 (1.82)*	0.04 (8.97)***	0.75	0.24	0.16	0.63

Note: (i) Figures in brackets are t values.
(ii) t values are '*', '**' and '***'significant at 5 per cent, 1 per cent and 10 per cent.

Appendix-21
Growth in Area, Production and Yield of Sugarcane in Punjab

Dependent Variables (Y)	Intercept (∝)				Slope (β)				R^2			
	1966-67 to 1980-81	1981-82 to 1995-96	1996-97 to 2014-15	1966-67 to 2014-15	1966-67 to 1980-81	1981-82 to 1995-96	1996-97 to 2014-15	1966-67 to 2014-15	1966-67 to 1980-81	1981-82 to 1995-96	1996-97 to 2014-15	1966-67 to 2014-15
Area	5.08	4.38	6.01	3.66	-0.04 (5.35)**	0.0008 (0.84)	-0.030 (4.26)**	0.009 (47.41)**	0.68	0.05	0.52	0.98
Production	6.24	6.31	6.74	6.30	0.001 (0.14)	0.009 (1.01)	-0.03 (4.05)**	0.002 (0.99)	0.002	0.07	0.49	0.02
Yield	8.06	7.70	8.69	8.40	0.04 (7.55)**	0.0001 (117.49)*	0.003 (0.20)	0.008 (5.99)**	0.81	0.99	0.002	0.43

Note: (i) Figures in brackets are t values.
(ii) t values are '*' and '**' significant at 1 per cent and 10 per cent.

Appendix-22
Growth in Area, Production and Yield of Cotton in Punjab

Dependent Variables (Y)	Intercept (∝)				Slope (β)				R^2			
	1966-67 to 1980-81	1981-82 to 1995-96	1996-97 to 2014-15	1966-67 to 2014-15	1966-67 to 1980-81	1981-82 to 1995-96	1996-97 to 2014-15	1966-67 to 2014-15	1966-67 to 1980-81	1981-82 to 1995-96	1996-97 to 2014-15	1966-67 to 2014-15
Area	5.91	6.37	6.76	6.30	0.03 (9.21)***	0.003 (0.49)	-0.01 (1.01)	0.007 (0.39)	0.87	0.02	0.08	0.003
Production	4.87	5.26	5.24	5.11	0.03 (8.47)**	0.04 (2.87)**	0.03 (2.18)**	0.01 (4.48)***	0.85	0.39	0.22	0.30
Yield	5.87	5.10	4.32	5.04	-0.001 (0.22)	0.04 (2.71)**	0.05 (3.64)***	0.002 (32.37)***	0.004	0.36	0.44	0.96

Note: (i) Figures in brackets are t values.
(ii) t values are significant at '**' 5 per cent and '***' per cent.

Appendix-23

Growth in Area, Production and Yield of Oilseeds in Punjab

Dependent Variables (Y)	Intercept (∝)				Slope (β)				R^2			
	1966-67 to 1980-81	1981-82 to 1995-96	1996-97 to 2014-15	1966-67 to 2014-15	1966-67 to 1980-81	1981-82 to 1995-96	1996-97 to 2014-15	1966-67 to 2014-15	1966-67 to 1980-81	1981-82 to 1995-96	1996-97 to 2014-15	1966-67 to 2014-15
Area	5.93	5.22	5.09	6.07	-0.02 (3.39)***	-0.007 (0.55)	-0.07 (9.92)***	-0.4 (18.02)***	0.47	0.02	0.85	0.87
Production	5.70	4.92	5.12	5.74	-0.02 (3.12)**	0.03 (1.44)	-0.05 (5.71)***	-0.2 (9.87)***	0.43	0.14	0.66	0.67
Yield	6.67	6.60	6.93	6.58	0.00005 (0.01)	0.03 (4.01)***	0.01 (3.04)**	0.01 (10.27)***	0.00	0.55	0.35	0.69

Note: (i) Figures in brackets are t values.
(ii) t values are significant at '***' 1 per cent.

Appendix-24
Growth in Area, Production of Non-Foodgrains in Haryana

Dependent Variables (Y)	Intercept (∝)				Slope (β)				R^2			
	1966-67 to 1980-81	1981-82 to 1995-96	1996-97 to 2014-15	1966-67 to 2014-15	1966-67 to 1980-81	1981-82 to 1995-96	1996-97 to 2014-15	1966-67 to 2014-15	1966-67 to 1980-81	1981-82 to 1995-96	1996-97 to 2014-15	1966-67 to 2014-15
Area	6.14	6.79	7.20	6.38	0.04 (1.78)**	0.02 (1.54)	-0.006 (1.67)	0.02 (7.19)***	0.20	0.16	0.14	0.52
Production	6.68	8.68	9.24	7.07	0.06 (2.74)**	0.04 (5.18)***	-0.0006 (0.11)	0.06 (11.46)***	0.37	0.67	0.001	0.74

Note: (i) Figures in brackets are t values.
(ii) t values are significant at '**' 10 per cent and '***'1 per cent.

Appendix-25
Growth in Area, Production and Yield of Sugarcane in Haryana

Dependent Variables (Y)	Intercept (∝)				Slope (β)				R^2			
	1966-67 to 1980-81	1981-82 to 1995-96	1996-97 to 2014-15	1966-67 to 2014-15	1966-67 to 1980-81	1981-82 to 1995-96	1996-97 to 2014-15	1966-67 to 2014-15	1966-67 to 1980-81	1981-82 to 1995-96	1996-97 to 2014-15	1966-67 to 2014-15
Area	4.98	4.92	5.15	5.07	0.003 (0.30)	-0.002 (0.32)	-0.03 (4.59)**	-0.07 (3.84)**	0.07	0.08	0.55	0.24
Production	6.42	6.31	6.79	6.36	-0.001 (0.11)	0.02 (2.83)**	-0.01 (1.83)**	0.006 (3.35)***	0.001	0.38	0.17	0.19
Yield	8.35	8.30	8.54	8.19	-0.004 (0.65)	0.02 (3.83)**	0.02 (9.33)***	0.01 (12.66)***	0.03	0.53	0.84	0.77

Note: (i) Figures in brackets are t values.
(ii) t values are significant at '***'1 per cent and '**' 10 per cent.

Appendix-26

Growth in Area, Production and Yield of Cotton in Haryana

Dependent Variables (Y)	Intercept (∝)				Slope (β)				R^2			
	1966-67 to 1980-81	1981-82 to 1995-96	1996-97 to 2014-15	1966-67 to 2014-15	1966-67 to 1980-81	1981-82 to 1995-96	1996-97 to 2014-15	1966-67 to 2014-15	1966-67 to 1980-81	1981-82 to 1995-96	1996-97 to 2014-15	1966-67 to 2014-15
Area	5.15	6.24	6.37	5.58	0.05 (2.0)**	-0.0008 (0.04)	-0.004 (0.94)	0.02 (5.25)***	0.24	0.00	0.05	0.37
Production	5.69	6.38	6.90	5.79	0.05 (10.44)***	0.06 (6.27)***	0.04 (4.6)***	0.04 (20.52)***	0.89	0.75	0.56	0.90
Yield	7.44	7.05	7.44	7.12	-0.01 (0.36)	0.06 (2.68)**	0.05 (4.64)**	0.02 (4.94)**	0.01	0.36	0.56	0.34

Note: (i) Figures in brackets are t values.
(ii) values are significant at '**' 5 per cent and '***' 1 per cent.

Appendix-27

Growth in Area, Production and Yield of Oilseeds in Haryana

Dependent Variables (Y)	Intercept (∝)				Slope (β)				R^2			
	1966-67 to 1980-81	1981-82 to 1995-96	1996-97 to 2014-15	1966-67 to 2014-15	1966-67 to 1980-81	1981-82 to 1995-96	1996-97 to 2014-15	1966-67 to 2014-15	1966-67 to 1980-81	1981-82 to 1995-96	1996-97 to 2014-15	1966-67 to 2014-15
Area	5.03	5.40	6.37	5.02	0.01 (0.75)	0.07 (7.88)***	-0.03 (0.54)	0.03 (10.82)***	0.04	0.02	0.83	0.71
Production	4.24	5.09	6.48	4.32	0.04 (2.00)**	0.11 (7.25)***	0.02 (2.17)**	0.06 (14.89)***	0.24	0.80	0.22	0.83
Yield	6.11	6.61	7.03	6.20	0.03 (2.02)**	0.04 (3.13)***	0.02 (2.80)*	0.02 (11.86)***	0.24	0.43	0.32	0.75

Note: (i) Figures in brackets are t values.
(ii) t values are significant at '***'1 per cent, '**' 10 per cent and '*' 5 per cent.

Appendix-28

Percentage Share of Agriculture in Total NSDP

Year	Haryana	Punjab
1966-67	56.26	51.65
1967-68	60.65	54.22
1968-69	55.00	52.75
1969-70	60.51	52.59
1970-71	64.43	58.09
1971-72	60.45	57.62
1972-73	58.15	55.66
1973-74	56.31	55.83
1974-75	56.22	56.91
1975-76	58.01	54.69
1976-77	55.75	53.27
1977-78	54.90	52.97
1978-79	54.31	52.83
1979-80	47.58	51.32
1980-81	51.64	49.77
1981-82	51.09	49.38
1982-83	50.23	49.46
1983-84	50.60	47.84
1984-85	49.96	49.68
1985-86	48.40	50.32
1986-87	45.86	48.38
1987-88	40.09	48.18
1988-89	46.47	47.86
1989-90	44.77	49.15
1990-91	45.06	47.63
1991-92	40.89	41.89
1992-93	40.57	40.94
1993-94	42.23	47.89
1994-95	42.34	47.60

Contd...

1995-96	39.42	45.74
1996-97	39.26	45.66
1997-98	35.35	42.09
1998-99	34.68	40.96
1999-2000	33.85	41.84
2000-01	32.43	41.41
2001-02	30.69	41.07
2002-03	28.96	38.67
2003-04	28.98	38.39
2004-05	27.56	37.89
2005-06	41.15	36.57
2006-07	41.03	36.44
2007-08	19.54	27.92
2008-09	19.47	26.93
2009-10	16.90	24.99
2010-11	16.61	23.74
2011-12	16.69	22.72
2012-13	22.35	31.99
2013-14	21.81	31.45
2014-15	20.26	29.37

Source: *Statistical Abstracts of Haryana and Punjab* for Various Years.

Appendix-29

Male Agricultural Labourers (1971-2015) (Lakhs)

Years	Punjab	Haryana
1971	7.67 (19.97)	3.97 (15.61)
1972	9.19 (23.79)	4.12 (16.64)
1973	9.29 (23.40)	4.27 (16.51)
1974	9.39 (23.03)	4.42 (16.39)
1975	9.49 (22.69)	4.56 (16.28)
1976	9.59 (22.36)	4.71 (16.18)
1977	9.7 (22.04)	4.86 (16.08)
1978	9.8 (21.74)	5.01 (16.00)
1979	9.9 (21.46)	5.15 (15.92)
1980	10 (21.18)	5.3 (15.84)
1981	10.47 (22.05)	5.28 (15.62)
1982	10.21 (20.67)	5.59 (15.70)
1983	10.31 (20.43)	5.74 (15.64)
1984	10.41 (20.21)	5.89 (15.58)
1985	10.51 (19.99)	6.04 (15.52)
1986	10.61 (19.78)	6.18 (15.47)
1987	10.71 (19.57)	6.33 (15.42)

Contd...

1988	10.82 (19.38)	6.48 (15.37)
1989	10.92 (19.19)	6.63 (15.32)
1990	11.02 (19.01)	6.77 (15.28)
1991	13.88 (23.80)	7.82 (18.35)
1992	11.22 (18.67)	7.07 (15.20)
1993	11.33 (18.51)	7.22 (15.16)
1994	11.43 (18.35)	7.36 (15.13)
1995	11.53 (18.20)	7.51 (15.09)
1996	11.63 (18.06)	7.66 (15.06)
1997	11.73 (17.92)	7.8 (15.03)
1998	11.83 (17.78)	7.95 (15.00)
1999	11.94 (17.65)	8.1 (14.97)
2000	12.04 (17.52)	8.25 (14.94)
2001	11.2 (15.94)	7.12 (12.46)
2002	12.24 (17.28)	8.54 (14.89)
2003	12.34 (17.16)	8.69 (14.86)
2004	12.45 (17.05)	8.84 (14.84)
2005	12.55 (16.94)	8.98 (14.82)

Contd...

2006	12.65 (16.83)	9.13 (14.79)
2007	12.75 (16.73)	9.28 (14.77)
2008	12.85 (16.63)	9.43 (14.75)
2009	12.95 (16.53)	9.57 (14.73)
2010	13.06 (16.43)	9.72 (14.71)
2011	12.39 (15.35)	10.41 (15.30)
2012	13.26 (16.25)	10.01 (14.68)
2013	13.36 (16.16)	10.16 (14.66)
2014	13.46 (16.08)	10.31 (14.64)

Source: *Census of India* for years 1971, 1981, 1991 and 2011 and for other years' values are estimated.
Note: In brackets percentages are of Male Agricultural Labourers Out of Total Male Workers.

Appendix-30

Money Wage Rate of Agricultural Labourers in Punjab and Haryana (Rs./Day)

Year	Punjab	Haryana
1970-71	6.39 (100)	6.64 (100)
1971-72	6.64 (104)	6.84 (103)
1972-73	7.06 (110)	7.12 (107)
1973-74	7.47 (117)	7.4 (111)
1974-75	8.75 (137)	8.58 (129)
1975-76	9.16 (143)	8.55 (129)
1976-77	10.01 (157)	8.75 (132)
1977-78	10.38 (162)	10.44 (157)
1978-79	10.73 (168)	11.17 (168)
1979-80	11.41 (179)	11.89 (179)
1980-81	12.23 (191)	12.41 (187)
1981-82	12.93 (202)	14.28 (215)
1982-83	13.62 (213)	16.14 (243)
1983-84	15.66 (245)	18.15 (273)
1984-85	18.13 (284)	19.35 (291)
1985-86	20 (313)	15 (226)

Contd...

1986-87	18.48 (289)	20.25 (305)
1987-88	23.67 (370)	22.01 (331)
1988-89	23.87 (374)	20 (301)
1989-90	28 (438)	20 (301)
1990-91	31.04 (486)	25 (377)
1991-92	37.7 (590)	27.75 (418)
1992-93	41.89 (656)	27.75 (418)
1993-94	47.81 (748)	45 (678)
1994-95	54.87 (859)	45 (678)
1995-96	55.58 (870)	54.45 (820)
1996-97	60.36 (945)	45 (678)
1997-98	65 (1017)	65 (979)
1998-99	63.3 (991)	69.2 (1042)
1999-2000	66.86 (1046)	73.64 (1109)
2000-01	81.62 (1277)	79.11 (1191)
2001-02	81.99 (1283)	80.78 (1217)
2002-03	84.05 (1315)	83.2 (1253)
2003-04	88.92 (1392)	90.22 (1359)

Contd...

2004-05	91.72 (1435)	92 (1386)
2005-06	95.09 (1488)	82.6 (1244)
2006-07	98.06 (1535)	112.65 (1697)
2007-08	102.22 (1600)	108.57 (1635)
2008-09	130.92 (2049)	148.89 (2242)
2009-10	139.8 (2188)	156.88 (2363)
2010-11	186.67 (2921)	199.21 (3000)
2011-12	205.5 (3216)	231.16 (3481)
2012-13	218.17 (3414)	266.5 (4014)
2013-14	304 (4757)	325.8 (4907)
2014-15	317.66 (4971)	371.5 (5595)

Note: (i) Money Wage Rate are of Male Agricultural Labourers at the ploughing season in the month of July.
(ii) The index is given in brackets.

Source: *Indian Labour Journal* (various issues), Labour Bureau, Government of India.

Appendix-31

Consumer Price Index (CPI) of Agricultural Labourers

Years	Punjab		Haryana	
	Base: 1960-61	Base: 1970-71	Base: 1960-61	Base: 1970-71
1970-71	211	100	211	100
1971-72	223	106	223	106
1972-73	250	118	250	118
1973-74	302	143	302	143
1974-75	375	178	375	178
1975-76	334	158	334	158
1976-77	326	155	326	155
1977-78	360	171	360	171
1978-79	383	182	383	182
1979-80	406	192	406	192
1980-81	486	230	486	230
1981-82	512	243	512	243
1982-83	520	246	520	246
1983-84	563	267	563	267
1984-85	621	294	621	294
1985-86	666	316	666	316
1986-87	656	311	656	311
1987-88	750	355	750	355
1988-89	885	419	885	419
1989-90	913	433	913	433
1990-91	1034	490	1034	490
1991-92	1159	549	1159	549
1992-93	1223	580	1223	580
1993-94	1446	685	1446	685
1994-95	1560	739	1560	739
1995-96	1588	753	1558	738
1996-97	1821	863	1821	863
1997-98	1949	924	1949	924

Contd...

1998-99	2167	1027	2167	1027
1999-2000	2183	1035	2183	1035
2000-01	2158	1023	2158	1023
2001-02	2218	1051	2218	1051
2002-03	2238	1061	2238	1061
2003-04	2329	1104	2329	1104
2004-05	2459	1165	2459	1165
2005-06	2641	1252	2641	1252
2006-07	2904	1376	2904	1376
2007-08	3130	1483	3130	1483
2008-09	3563	1689	3563	1689
2009-10	4290	2033	4290	2033
2010-11	4515	2140	4515	2140
2011-12	4955	2348	4955	2348
2012-13	5327	2525	5327	2525
2013-2014	5820	2758	5820	2758
2014-2015	6134	2907	6134	2907

Note: For the common base year we used the linking factor of CPI (Agricultural Labourers) given in *Indian Labour Journal*, February 1996.
Source: *Statistical Abstracts of India*.

Appendix-32
Real Wage Rate of Male Agricultural Labourers (Rs./Day)

Year	Punjab	Haryana
1970-71	3.03 (100)	3.15 (100)
1971-72	3.15 (104)	3.24 (103)
1972-73	3.35 (110)	3.37 (107)
1973-74	3.54 (117)	3.51 (111)
1974-75	4.15 (137)	4.07 (129)
1975-76	4.34 (143)	4.05 (129)
1976-77	4.74 (157)	4.15 (132)
1977-78	4.92 (162)	4.95 (157)
1978-79	5.09 (168)	5.29 (168)
1979-80	5.41 (179)	5.64 (179)
1980-81	5.8 (191)	5.88 (187)
1981-82	6.13 (202)	6.77 (215)
1982-83	6.45 (213)	7.65 (243)
1983-84	7.42 (245)	8.6 (273)
1984-85	8.59 (284)	9.17 (291)
1985-86	9.48 (313)	7.11 (226)
1986-87	8.76 (289)	9.6 (305)

Contd...

1987-88	11.22 (370)	10.43 (331)
1988-89	11.31 (374)	9.48 (301)
1989-90	13.27 (438)	9.48 (301)
1990-91	14.71 (486)	11.85 (377)
1991-92	17.87 (590)	13.15 (418)
1992-93	19.85 (656)	13.15 (418)
1993-94	22.66 (748)	21.33 (678)
1994-95	26 (859)	21.33 (678)
1995-96	26.34 (870)	25.81 (820)
1996-97	28.61 (945)	21.33 (678)
1997-98	30.81 (1017)	30.81 (979)
1998-99	30 (991)	32.8 (1042)
1999-2000	31.69 (1046)	34.9 (1109)
2000-01	38.68 (1277)	37.49 (1191)
2001-02	38.86 (1283)	38.28 (1217)
2002-03	39.83 (1315)	39.43 (1253)
2003-04	42.14 (1392)	42.76 (1359)
2004-05	43.47 (1435)	43.6 (1386)

Contd...

2005-06	45.07 (1488)	39.15 (1244)
2006-07	46.47 (1535)	53.39 (1697)
2007-08	48.45 (1600)	51.45 (1635)
2008-09	62.05 (2049)	70.56 (2242)
2009-10	66.26 (2188)	74.35 (2363)
2010-11	88.47 (2921)	94.41 (3000)
2011-12	97.39 (3216)	109.55 (3481)
2012-13	103.4 (3414)	126.3 (4014)
2013-14	144.08 (4757)	154.41 (4907)
2014-15	151 (4986)	176.07 (5595)

Note: Real Wage Rate was calculated on the basis of CPI (Food) for Agricultural Labourers for the year 1970-71.

Appendix-33

Development Indicators: Socio-Economic and Agriculture: Punjab and Haryana

Sr. No.	Indicators	Unit	Year	Punjab	Haryana
1.	Total Area	Sq. Km.	2011	50362	44212
2.	Total Area: Percentage Share of India	Percentage	2011	1.53	1.34
3.	Population	Crores	2011	2.77	2.53
4.	Population: Percentage Share of India	Percentage	2011	2.29	2.09
5.	Population Density	Per Sq. Km.	2011	551	573
6.	Number of Villages	Number	2011	12581	6841
7.	Percentage of Rural Population Out of Total Population of a State	Percentage	2011	62.52	65.12
8.	Percentage of Scheduled Caste Population Out of Total Population of a State	Percentage	2011	31.94	20.17
9.	Percentage of Scheduled Caste Population Out of India's Total Scheduled Caste Population	Percentage	2011	4.39	2.54
10.	Annual Population Growth Rate	Percentage	1971-2011	1.81	1.88
11.	Crude Birth Rate	Percentage	2015	15.2	22.4
12.	Crude Death Rate	Percentage	2015	6.2	6.6
13.	Percentage of Stunted Children in Rural Areas	Percentage	2015-16	24.5	34.3
14.	Number of Anganwaris in Rural Areas	Number	2011	24494	24027
15.	Percentage of Households Having Access to Safe Drinking Water in Rural Area	Percentage	2015-16	99.0	94.3

Contd...

16.	Percentage of Households Having Access to Toilets in Rural Area	Percentage	2015-16	79.1	77.4
17.	Percentage of Electrified Houses in Rural Area	Percentage	2015-16	99.6	98.3
18.	Hunger Index	Rank	2009	1	5
19.	Literacy Rate	Percentage	2011	75.8	75.6
20.	Rural Literacy Rate	Percentage	2011	71.4	71.4
21.	Primary Schools in Rural Areas	Number	2015-16	12430	8423
22.	Matric Schools in Rural Areas	Number	2015-16	1722	1326
23.	Senior Secondary Schools in Rural Areas	Number	2015-16	1390	1626
24.	Total Cropped Area	('000 Hectares)	2014-15	7857	6536
25.	Net Sown Area	('000 Hectares)	2014-15	4118	3522
26.	Percentage of Net Area Sown to Total Geographical Area	Percentage	2015-16	81.83	80.57
27.	Net Sown Area as Percentage Share of India	Percentage	2014-15	2.91	2.49
28.	Net Irrigated Area	('000 Hectares)	2014-15	4143	2931
29.	Percentage of Net Irrigated Area Out of Net Sown Area	Percentage	2014-15	100	83.21
30.	Gross Irrigated Area	('000 Hectares)	2014-15	7757	5824
31.	Percentage of Gross Irrigated Area Out of Total Cropped Area	Percentage	2014-15	98.9	89.10
32.	Irrigation Intensity		2014-15	187	198
33.	Average Size of Operational Land Holdings	Hectares	2010-11	3.77	2.25

Contd...

34.	Gini Coefficient of Ownership of Land	Hectares	1970-71	0.50	0.52
35.	Coefficient of Variation of Ownership of Land	Hectares	1970-71	1.03	1.11
36.	Gini Coefficient of Ownership of Land	Hectares	2010-11	0.33	0.18
37.	Coefficient of Variation of Ownership of Land	Hectares	2010-11	0.61	0.33
38.	Percentage of Area Under Vegetables Cultivation Out of Total Cropped Area	Percentage	2014-15	1.67	0.78
39.	Percentage of Area Under High-Yielding Varieties	Percentage	2014-15	99.9	77.5
40.	Percentage Share in India's Milk Production	Percentage	2014-15	7.07	5.40
41.	Consumption of Fertilisers Per Hectare (Kg)	Kilograms	2015-16	213	199
42.	Percentage of Agricultural Workers (Cultivators + Agricultural Labourers) in Total Workers	Percentage	2011	35.59	44.96
43.	Percentage of Female Agricultural Labourers Out of Total Agricultural Labourers	Percentage	2011	21.97	31.86
44.	Average Number of Village Served Per Regulated Markets (Agricultural)	Number	2015-16	81	64
45.	Commercial Bank Branches in Rural Areas (Per 1000 Persons)	Number	2015	6210	4544
46.	Per Cultivator Loan Disbursed by Commercial Bank Branches	Rupees	2011-12	19538	37508

Contd...

47.	Per Cultivator Loan Disbursed by PACs	Rupees	2011-12	30328	26341
48.	Surfaced Road Length Per Village	Km per Village	2015-16	7.45	6.12
49.	Infrastructure Index	Rank (in Ascending Order)	2012	4	9
50.	Per Capita Income at Current Prices	Rupees	2015-16	119261	162034
51.	Per Capita Income at Constant Prices (2011-12)	Rupees	2015-16	99372	133591
52.	Percentage of Persons Below Poverty Line in Rural Area	Percentage	2011-12	7.7	11.6
53.	MPCE on Food Items in Rural Areas	Rupees	2009-10	795.01	815.20
54.	MPCE on Non-Food Items in Rural Areas	Rupees	2009-10	1648.92	1509.91

Source: Sr. No. (1) *Statistical Abstracts of Punjab and Haryana 2015-16.* (2) *Statistical Abstracts of Punjab, Haryana and India, 2015-16.* (3) *Census of India, 2011.* (4) *Census of India, 2011.* (5) *Census of India, 2011.* (6) *Census of India, 2011.* (7) *Census of India, 2011.* (8) *Census of India, 2011.* (9) *Census of India, 2011.* (10) *Calculated from Census of India, 2011.* (11) *National Family Health Survey, 2015-16.* (12) *National Family Health Survey, 2015-16.* (13) *National Family Health Survey, 2015-16.* (14) Kumar, S., Banerjee, S. (2015) "Integrated Child Development Services (ICDS) Programme in the Context of Urban Poor and Slum Dwellers in India: Exploring Challenges and Opportunities", *Indian Journal of Public Administration, Vol. LXI, No. 1, January-March 2015.* (15) *National Family Health Survey, 2015-16.* (16) *National Family Health Survey, 2015-16.* (17) *National Family Health Survey, 2015-16.* (18) *Indian State Hunger Index: Comparison of Hunger Across States, February 2009,* IFPRI. (19) *Statistical Abstracts of Punjab and Haryana 2015-16.* (20) *Statistical Abstracts of Punjab and Haryana 2015-16.* (21) *Elementary Education in India, State Report Cards, 2015-16,* NUEPA, MHRD, New Delhi. (22) *Elementary Education in India, State Report Cards, 2015-16,* NUEPA, MHRD, New Delhi. (23) *Elementary Education in India, State Report Cards, 2015-16,* NUEPA, MHRD, New Delhi. (24) *Statistical Abstracts of Punjab and Haryana 2015-16.* (25) *Statistical Abstracts of Punjab and Haryana 2015-16.* (26) *Statistical Abstracts of Punjab and Haryana 2015-16.* (27) *Statistical Abstracts of Punjab, Haryana and India, 2015-16.* (28) *Statistical Abstracts of Punjab and Haryana 2015-16.* (29) *Statistical Abstracts of Punjab and Haryana 2015-16.* (30) *Statistical Abstracts of Punjab, Haryana and India 2015-16.* (31) *Statistical Abstracts of Punjab and Haryana 2015-16.* (32) Calculated by dividing Gross Irrigated Area by Net Irrigated Area and then Multiplying by 100. (33) Agricultural Census, 2010-11. (34) Calculated by Using *Agricultural Census, 1970.* (35) Calculated by Using *Agricultural Census, 1970.* (36) Calculated by Using *Agricultural Census, 2010-11.* (37) Calculated by Using *Agricultural Census, 2010-11.* (38) *Statistical Abstracts of Punjab and Haryana 2015-16.* (39) *Statistical Abstracts of Punjab and Haryana 2015-16.* (40) *Statistical Abstracts of Punjab, Haryana and India 2015-16.* (41) *Statistical Abstracts of Punjab and Haryana 2015-16.* (42) *Census of India, 2011.* (43) *Census of India, 2011.* (44) *Statistical Abstracts of Punjab and Haryana 2015-16.* (45) *Statistical Abstracts of Punjab and Haryana 2015-16.* (46) *Agricultural Census, 2011-12.* (47) *Agricultural Census, 2011-12.* (48) *Statistical Abstracts of Punjab and Haryana 2015-16.* (49) *Refining State Level Comparison in India, Working Paper Series 2012,* Planning Commission, Government of India. (50) *Statistical Abstracts of Punjab and Haryana 2015-16.* (51) *Statistical Abstracts of Punjab and Haryana 2015-16.* (52) *Report of the Expert Group to Review the Methodology for Measurement of Poverty*, Planning Commission, Government of India, June 2014. (53) *Household Consumption of Various Goods and Services in India, NSS 66th Round, July 2009-10,* Government of India. (54) *Household Consumption of Various Goods and Services in India, NSS 66th Round, July 2009-10,* Government of India.

Bibliography

Acharya, S.S. (1973): "Green Revolution and Farm Employment". *Indian Journal of Agricultural Economics*, Vol. 28, No. 3, pp. 30-45.

Acharya, S.S. and Jogi, R.L. (2004): *Farm Input Subsidies in Indian Agriculture*. Institute for Development Studies, Jaipur.

Aggarwal, Bina (1983): *Mechanisation in Indian Agriculture: An Analytical Study Based on the Punjab.* Allied Publishers, New Delhi.

Ahluwalia, M.S. (1978): "Rural Poverty and Agricultural Performance in India". *Journal of Development Studies,* Vol. 14, No. 13, pp. 298-323.

Ahuja and Rajdeep (1985): *Agricultural Development in Punjab.* Allied Publishers, Delhi.

Alagh, Y.K., Bhalla, G.S. and Kashyap, S.P. (1980): *Structural Analysis of Gujarat, Punjab and Haryana Economies.* Allied Publishers, New Delhi.

Aulakh, Harbans Singh (1983): *Changing Foodgrain Market Structure in Punjab.* B.R. Publishing Corporation, Delhi.

Banga, Indu (1978): *The Agrarian System of the Sikhs (Late 18th Early 19th Century).* Manohar Publications, Delhi.

Bansil, P.C. (1981): *Agricultural Problems of India.* Oxford & IBM Publishing Co., New Delhi.

Bardhan, Pranab (1970): "Green Revolution and Agricultural Labourers". *Economic and Political Weekly,* Vol. 5, No. 29-31, pp. 1239-1246.

Bardhan, Pranab (2003): *Poverty, Agrarian Structure and Political Economy in India.* Oxford University Press, New Delhi.

Bardhan, Pranab (2010): *Awakening Giants: Feet of Clay,* Oxford University Press, Delhi.

Bardhan, Pranab and Rudra, Ashok (1981): "Terms and Conditions of Labour Contracts in Agriculture, Results of a Survey in West Bengal, 1979". *Bulletin of the Oxford University of Institute of Economics and Statistics*, Vol. 43, No. 1, pp. 89-111.

Bardhan, Pranab K. (1984): *Land, Labour and Rural Poverty*. Oxford University Press, Delhi.

Bhalla, G.S. and Singh, Gurmail (2001): *Indian Agriculture: Four Decades of Development*. Sage Publications, New Delhi.

Bhalla, G.S. (1974): *Changing Agrarian Structure of India: A Study of the Impact of Green Revolution in Haryana*. Meenakshi Prakashan, Meerut.

Bhalla, G.S. (2012): *Economic Liberalisation and Indian Agriculture*. Sage Publications, New Delhi.

Bhalla, G.S. and Chadha, G.K. (1980): *Structure of Institutional Set Up of Rural Punjab in the Year 2000s* (Mimeo). FAO, Rome.

Bhalla, G.S. and Tyagi, G.S. (1989): *Patterns in Indian Agricultural Development: A District Level Study*. Institute for Studies in Industrial Development, New Delhi.

Bhalla, Sheila (1978): "New Relations of Production in Haryana". *Economic and Political Weekly*, Vol. 11, No. 13, pp. A23-A30.

Bhalla, Sheila (1979): "Real Wage Rates of Agricultural Labourers in Punjab, 1961-1977, A Preliminary Analysis". *Economic and Political Weekly*, Vol. 14, No. 26, pp. A57-A68.

Bhalla, Sheila (1981): "Islands of Growth: A Note of Haryana's Experience and Some Possible Implications". *Economic and Political Weekly*, Vol. 16, No. 23, pp. 1020-1030.

Bhalla, Sheila (1977), "Changes in Acreage and Tenure Structure of Land Holdings in Haryana, 1962-72". *Economic and Political Weekly*, Vol. 12, No. 13, pp. A2-A15.

Bhalla, Sheila (1979), "Real Wage Rates of Agricultural Labourers in Punjab, 1961-77—A Preliminary Analysis". *Economic and Political Weekly*, Vol. 14, No. 26, pp. A57-A68.

Bhangoo, Singh, Kesar, Singh, Lakhwinder and Sharma, Rakesh (2016): *Agrarian Distress and Farmers' Suicides in North India*. Routledge Publication, Delhi.

Bhattacharya, Pranab and Ahdil Majid Jr. (1976): "Impact of Green Revolution on Output, Cost and Income of Small and Big Farmers". *Economic and Political Weekly*, Vol. 1, No. 52, pp. A147-A150.

Bhatty, I.Z. (1974): "Inequality and Poverty in Rural India", in *Poverty and Income Distribution in India* (ed.). Statistical Publishing Society, Calcutta.

Billings, Martin H and Singh, Arjan (1969): "Labour and the Green Revolution: The Experience of Punjab". *Economic and Political Weekly*, Vol. 4, No. 52, pp. A221-224.

Billings, Martin H and Singh, Arjan (1970): "Mechanisation and the Wheat Revolution: Effects on Female Labour in Punjab". *Economic and Political Weekly*, Vol. 5, No. 52, pp. A169-174.

Binswanger, H.P. (1978): *The Economics of Tractors in South Asia.* Agricultural Development Council, New York.

Biswanger, Hans P. and Rosenzuceig, Mark, K. (1984): *Contractual Arrangements, Employment and Wages in Rural Labour Markets in Asia.* Yale University Press, Haven.

Blyn, George (1983): "Income Distribution Among Haryana & Punjab Cultivators—1968/69-1975/76". *Indian Economic Review*, Vol. 18, No. 2, pp. 199-224.

Brass, Tom (1990): "Class Struggle and Deproletarianisation of Agricultural Labour in Haryana (India)". *The Journal of Peasant Studies*, Vol. 18, No. 1, pp. 36-67.

Brown, Dorris D. (1971): *Agricultural Developments in India's Districts.* Harvard University Press, Cambridge.

Chadha, G.K. (1979): *Production Gains of New Agricultural Technology.* Punjab University Publication Bureau, Chandigarh.

Chadha, G.K. (1986): *The State and Rural Economic Transformation: The Case Study of Punjab (1950-85).* Sage Publications, New Delhi.

Chand, Krishan (2002): *Migrant Labour and the Trade Union Movement in Punjab: A Case Study of the Sugar Industry*, CRRID, Chandigarh.

Chand, Ramesh (2016): "Why Doubling Farmers Income by 2022 is Possible". *Indian Express*, April 4, 2016.

Darling, Malcolum Lyall (1934): *Wisdom and Waste in the Panjab Village.* Oxford University Press, London.

Day, Richard P. and Singh, Inderjit (1977): *Economic Development as an Adaptive Process: The Green Revolution in the Indian Punjab.* Cambridge University Press, Cambridge.

Devi, S. Mahendra (2016): *Poverty and Employment: Roles of Agriculture and Non-Agriculture.* V.V. Giri Memorial Lecture, 58th Annual Conference, Indian Society of Labour Economics, IIT, Guwahati.

Dhawan, B.D. (1982): *Development of Tubewell Irrigation in India.* Agricole Publishing Academy, New Delhi.

Frankel, Francine (1971): *India's Green Revolution: Economic Gains and Political Costs.* Princeton University Press, New Jersey.

Ghosh, R.N. (1977): *Agriculture in Economic Development with Special Reference to Punjab.* Vikas Publishing House, New Delhi.

Ghuman, Ranjit Singh, Singh, Inderjeet and Singh, Lakhwinder (2007): *Status of Local Agricultural Labour in Punjab.* Department of Economics, Punjabi University, Patiala.

Gill, Khem Singh (1993): *A Growing Agricultural Economy: Technological Changes, Constraints and Sustainability.* Oxford University Press, Delhi.

Gill, S.S. (1994): *Economic Development and Structural Change in Punjab: Some Policy Issues.* Euho Publishers, Ludhiana.

Gill, Sucha Singh (1994): *Economic Development and Structural Change in Punjab: Some Policy Issues.* Principal Iqbal Singh Memorial Lecture, Ludhiana.

Gosal, G.S. and Krishan, Gopal (1984): *Regional Disparities in Levels of Socio-Economic Development in Punjab.* Vishal Publications, Kurukshetra.

Gough, Kathleen (1981): *Rural Society in South East India.* Cambridge University Press, Cambridge.

Government of Punjab (2006): *Agricultural and Rural Development of Punjab: Transforming from Crisis to Growth*, The Punjab State Farmers' Commission, Mohali.

Grewal, S.S. and Rangi, P.S. (1983): "An Analytical Study of Growth of Punjab Agriculture". *Indian Journal of Agricultural Economics*, Vol. 38, No. 4, pp. 509-519.

Gupta, D.P. and Srigari, K.K. (1980): *Agricultural Development in Punjab.* Agricole Publishing, Delhi.

Jaynes, Gerald David (1982): "Production and Distribution in Agrarian Economies". *Oxford Economic Papers*, Vol. 34, No. 2, pp. 346-367.

Jodhka, Surinder S. (2014): "Emerging Ruralities: Revisiting Village Life and Agrarian Change in Haryana", *Economic and Political Weekly*, Vol. 49 (26-27), pp. 5-17.

Jodhka, Surinder S.: Emerging Ruralities: Revisting Village Life and Agrarian Change in Haryana. *Economic and Plitical Weekly*, Vol. 49, No. 26, 2014 and H.S. Shergill and Varinder Sharma: *Milk Production As Supplementary Source of Income for Rural Scheduled Caste Households in Punjab: Current Status and Strategy for Expansion and Viability.* 2018, Institute for Development and Communication.

Johl, S.S. (1975): "Gains of Green Revolution: How They Have Been Shared in Punjab". *Journal of Development Studies*, Vol. 11, No. 3, pp. 178-189.

Johl, S.S. (2002): *Future of Agriculture in Punjab.* CRRID, Chandigarh.

Joshi, P.K., Gulati, Ashok and Gemnings Jr., Ralph (2007): *Agricultural Diversification and Small Holders in South Asia.* Academic Foundation, Delhi.

Kahlon, A.S. (1984): *Modernisation of Punjab Agriculture.* Allied Publishers, New Delhi.

Kapila, Uma (1995): *Indian Economy Since Independence (1946-1995).* Academic Foundation, Delhi.

Khurana, M.R. (2018): *Dynamics of Rural Transformation in India* (ed.). Studera Press, New Delhi.

Kumar, A. Ganesh, Roy. Davesh and Gulati, Ashok (2011): *Liberalising Foodgrain Markets: Experiences Impact and Lessons from South Asia.* Oxford University Press, Delhi.

Kumar, Pramod, Sharma, Varinder and Sharma, S.L. (2006): *Suicides in Rural Punjab.* Institute for Development and Communication (IDC), Chandigarh.

Kundu, T.R., Kaushik, S. and Sharma, C.B. (1998): *A Study of Rural Indebtedness Among the Farmers in Haryana.* Department of Economics, Kurukshetra University (Haryana).

Majidi, A. and Talib, B.D. (1976): "The Small Farmers of Punjab". *Economic and Political Weekly*, Vol. 11, No. 26, pp. A42-A46.

Mathew, Rodell et al. (2009): "Satellite Based Estimates of Ground Water Depletion in India". *Nature*, Vol. 460, pp. 999-1002.

Mellor, W. John and Lete, J. Uma (1973): "Growth Linkages of the New Foodgrain Technologies". *Indian Journal of Agricultural Economics*, Vol. 28, No. 1, pp. 35-55.

Minhas, B.S. and Vaidyanathan, A. (1965): "Growth of Crop Output in India, 1951-1954". *Journal of the Indian Society of Agricultural Statistics*, Vol. 16, No. 1, pp. 1-50.

Mitra, Ashok (1978): *India's Population: Aspects of Quality and Control.* Abhinav Publications, New Delhi.

Mohan, R., Jhay D. and Everson, R.E. (1973): "The Indian Agricultural Research System". *Economic and Political Weekly*, Vol. 8, No. 13, pp. A23.

Nair, Kusum (1961): *Blossoms in the Dust.* Gerald Rockwork and Company, London.

Omvedt, Gail (1981): "Capitalist Agriculture and Rural Classes in India". *Economic and Political Weekly*, Vol. 16, No. 52, pp. A140-A159.

Omvedt, Gail (1982): *Land, Caste and Politics in Indian States.* Wild Publications, Michigan.

Oomen, T.K. (1971): "Green Revolution and Agrarian Conflict". *Economic and Political Weekly*, Vol. 6, No. 26, pp. A99-A103.

Panagariya, Arvind and Rao, Govind M. (2015): *The Making of Miracles in Indian States: Andhra Pradesh, Bihar and Gujarat.* Oxford University Press, Oxford.

Pandit, M.L. (1978): "Some Less Known Factors Behind Recent Industrial Change in Punjab and Haryana." *Economic and Political Weekly*, Vol. 13, No. 17, p. 1937.

Patnaik, Ultra (1987): *Peasant Class Differentiations: A Study in Method with Reference to Haryana.* Oxford University Press, Delhi.

Paul, Satya (1989): "Green Revolution and Income Distribution Among Farm Families in Haryana, 1969-70 to 1982-83". *Economic and Political Weekly*, Vol. 24, Nos. 51-52, pp. A154-A158.

Punjab State Farmers Commission: *Organic Farming—Its Necessity: How Far It Can Go?*, 2006.

Raj, K.N. (1985): *An Essay on the Commercialisation of Indian Agriculture* (ed.). Oxford University Press, Delhi.

Randhawa, M.S. (1974): *Green Revolution: A Case Study of Punjab.* Vikas Publishing House, New Delhi.

Resuming Punjab's Prosperity: *The Opportunities and Challenges Ahead*, WB, 2004.

Reynolds, Lloyd. G. (1975): *Agriculture in Development Theory* (ed.). Yale University Press, New Haven, C.T.

Rudra, Ashok (1971): "Employment Patterns in Large Farms of Punjab". *Economic and Political Weekly*, Vol. 6, No. 26, pp. A89-A94.

Rudra, Ashok (1971): *Indian Agricultural Economics: Myths and Realities.* Allied Publishers, New Delhi.

Sarkar, Suman (1986): "India's Agricultural Development: An Alternative Path". *Economic and Political Weekly*, Vol. 21, No. 19, pp. 825-838.

Schultz, Theodore. W. (1984): *Transforming Traditional Agriculture.* New Haven, Yale University Press, New Haven.

Sen, Bhowani (1962): *Evolution of Agrarian Relations in India.* People's Publishing House, Delhi.

Sharma, Varinder (1995): *Agricultural Development and Character of Industralisation: A Comparative Study of Punjab and Haryana.* M. Phil Dissertation (Unpublished), Department of Economics, Panjab University, Chandigarh.

Sharma, Varinder (2016): *Farm Workers of Punjab.* LG Publications, New Delhi.

Sharma, Varinder (2016): *Growth of Small Farmers in Punjab (1970-2011).* Institute for Development and Communication (IDC), Chandigarh.

Sharma, Varinder (2017): "Growth of Wage Labour in Punjab Agriculture: 1901 to 2011". *Panjab University Research Journal (Arts)*, Vol. 54, No. 1, pp. 45-54.

Shenoli, P.V. (1975): *Agricultural Development in India.* Vikas Publishing House, New Delhi.

Shergill (2013): *Why Area Under Rice Will Not Be Reduced in Punjab.* Institute for Development and Communication, Chandigarh.

Shergill, H.S. (1982): *Land Market Transactions and the Process of Growth and Decay of Peasant Farms: A Case Study of Sangrur District in Punjab.* Department of Economics, Panjab University, Chandigarh.

Shergill, H.S. (1998): *Rural Credit and Indebtedness in Punjab.* Institute for Development and Communication (IDC), Chandigarh.

Shergill, H.S. (2005): *Wheat and Paddy Cultivation and the Question of Optimal Cropping Pattern for Punjab.* Institute for Development and Communication (IDC), Chandigarh.

Shergill, H.S. (2006): *Diversification of Cropping Pattern: A Re-examination.* Institute for Development and Communication (IDC), Chandigarh.

Shergill, H.S. (2010): *Growth of Farm Debt in Punjab (1997-2008).* Institute for Development and Communication, Chandigarh.

Shergill, H.S. (2011): *Role of Cooperative Institutions in Agricultural Credit in Punjab.* Institute for Development and Communication, Chandigarh.

Shergill, H.S. (2012): *Strategy for Sustainable Expansion of Foodgrains Production in Punjab.* Institute for Development and Communication (IDC), Chandigarh.

Shergill, H.S. (2012): *Strategy for Sustainable Expansion of Foodgrains Production in Punjab.* Institute for Development and Communication, Chandigarh.

Shergill, H.S., Sucha Singh and Gurmail Singh (2011): *Understanding North-Western Indian Economy*, Serial Publications, New Delhi.

Shwartzberg, Joseph E. (1978): *A Historical Atlas of South Asia.* University of Chicago Press, Chicago.

Sidhu, M.S. et al (2009): "Sources, Replacement and Management of Paddy Seeds by Farmers in Punjab". *Agricultural Economics Research Review*, Vol. 22, No. 2, pp. 323-328.

Singer, Milton and Cohn, Bernard (1968): *Structure and Change in Indian Society.* Aldine Publishing Company, Chicago.

Singh, Baldev (1973): *Capital Formation in Haryana Agriculture.* Ph.D. Thesis, Kurukshetra University, Kurukshetra.

Singh, Karam, Sukhpal Singh and H.S. Kingra (2007): *State of Farmers Who Left Farming in Punjab.* The Punjab State Farmers' Commission (Government of Punjab).

Singh, Master Hari (1980): *Agricultural Workers' Struggle in Punjab.* People's Publishing House, New Delhi.

Singh, Pritam (1983): *Emerging Pattern of Punjab Economy.* Sterling Publishers, New Delhi.

Singh, Sukhpal and Bhogal, Shruti (2014): "Depeasantisation in Punjab: Status of Farming Who Left Farming". *Current Science*, Vol. 106, No. 10, pp. 1364-1381.

Singh, Sukhpal, Shruti Bhogal and Randeep Singh (2014): "Magnitude and Determinants of Indebtedness Among Farmers in Punjab", *Indian Journal of Agricultural Economics*, Vol. 69, No. 2, pp. 243-256.

Thorner, Daniel (1955): *The Agrarian Prospects in India.* Delhi School of Economics, University Press, Delhi.

Index